Reinforced Concrete and the Modernization
of American Building, 1900–1930

JOHNS HOPKINS STUDIES
IN THE HISTORY OF TECHNOLOGY

Merritt Roe Smith, Series Editor

Reinforced Concrete and the Modernization of American Building, 1900–1930

AMY E. SLATON

The Johns Hopkins University Press
BALTIMORE AND LONDON

© 2001 The Johns Hopkins University Press
All rights reserved. Published 2001
Printed in the United States of America on acid-free paper
9 8 7 6 5 4 3 2 1

The Johns Hopkins University Press
2715 North Charles Street
Baltimore, Maryland 21218-4363
www.press.jhu.edu

Frontispiece: Cover of catalog produced by the Aberthaw Company, Boston, 1926. The firm, founded in 1894, specialized in the construction of concrete buildings for industry. *ACC Archives*

Library of Congress Cataloging-in-Publication Data

  Slaton, Amy E., 1957–
    Reinforced concrete and the modernization of American building, 1900–1930 / Amy E. Slaton.
      p. cm. — (Johns Hopkins studies in the history of technology)
    Includes bibliographical references and index.
    ISBN 0-8018-6559-X (alk. paper)
    1. Reinforced concrete construction—United States—History. I. Title.
  II. Series.
TA683 .S575 2001
721'.04454'0973—dc21

                                                           00-044363

A catalog record for this book is available from the British Library.

Contents

Preface ix

INTRODUCTION
Science and Commerce: Scenes from a Marriage 1

CHAPTER ONE
Concrete Testing: The Academics at Work 20

CHAPTER TWO
Science on Site: The Field-Testing and Regulation of Concrete Construction 62

CHAPTER THREE
Science and the "Fair Deal": Standards, Specifications, and Commercial Ambition 95

CHAPTER FOUR
The Business of Building: Technological and Managerial Techniques in Concrete Construction 127

CHAPTER FIVE
What "Modern" Meant: Reinforced Concrete and the Social History of Functionalist Design 168

Conclusion 188

Notes 195
Bibliographic Essay 237
Index 249

Preface

Like all histories, this one follows a certain political agenda, and it is perhaps a particularly thinly veiled agenda: to demonstrate that fiscal, aesthetic, and especially scientific modernity in America after 1900 are best understood when examined through the lens of labor history, with its emphases on class, ethnic, and gender relations. Cultural developments that unfolded in this country between roughly 1900 and 1930, including the creation of important new visual forms and categories of intellectual accomplishment, had repercussions for many people outside the professional circles that generated those changes. Study of the physical artifacts of modernizing trends affords us a chance to locate these repercussions. The artifact of modernity featured here is one of little glamour but significant cultural influence: the functionalist reinforced-concrete factory building that came to dominate American industrial landscapes in the twentieth century. I explore how the labor of creating concrete buildings shaped the lives of materials scientists, engineers, builders, and rank-and-file construction workers in vastly different ways.

In seeing these buildings as reflections of significant social upheaval, I treat episodes of commercial, architectural, and technological change as "second-order" events following from increasingly hierarchical organizations of management and labor in this country. This stress on social mechanisms of modernization undermines two conventional historical narratives about early-twentieth-century America with which I first approached this subject: that it was a search for productive efficiency that exclusively or even primarily drove managerial innovation in this era; and that modernist designers instilled Americans' taste for modern design.[1] The concrete factory buildings called forth a more integrated story of professional enterprise than either of these narratives suggested. First, I found that businessmen and architects had a great deal to do with one another in this period. A glance at the journals of either profession shows a system of mutual address and, it becomes clear, influence. And of course, that most celebrated group of Progressive Era players, technical experts, made themselves indispensable to many such professionals by offering unprecedented services in industrial, academic, and government venues. All of these groups of accomplished practitioners came into repeated contact with one another and were obviously "getting along" on some level. Vitally, I came to see in writing this book that connections among all of these occupations went far deeper than relationships of client and provider to a network of cultural and social affiliations. Educators, engineers, industrialists, and builders expressed values in the conduct of business that made manifest shared ideologies of class, ethnic, and gender distinction. Many held consonant ideas about the optimal organization of American labor. If at moments I doubted the "concreteness" of such social commonalities, the discovery that purveyors of cement, successful architects, and academic concrete experts attended not just the same professional meetings but at times the same churches and country clubs spurred on this inquiry.[2]

To articulate the conceptual and social connections among these men—and these were almost universally male purviews—I turned to methods of science studies that focus on the development and control of knowledge systems. The idea is to approach technical work—as experienced in the upper and lower reaches of the economy—as agglomerations of knowledge (on whatever level) and therefore of opportunity (or its absence). In industrial and scientific fields after 1900, bodies of information and techniques were things to be guarded and exchanged with great self-consciousness. The ostensible possession of objective scientific skills or data, and even the association of objectivity with science, was a powerful social and commercial claim. The development of concrete as a commercial building medium involved many definitions and distributions of technical knowledge, bringing not just scientific practitioners but their industrial colleagues into elevated occupational and cultural standing. Rank-and-file workers on the modern concrete building site, from heavy laborers to skilled members of established building trades, achieved little in these distributions of emergent expertise.

It may seem odd to imbue such undistinguished, and indistinguishable, objects as concrete factory buildings with such intellectual and political significance, but if they do not have beauty or outward complexity to recommend them as important cultural markers, they do have sheer numbers. The origins of this narrative rest in the many hours I have spent riding trains between Boston, New York, and Washington, the corridor along which American manufacturing grew most dramatically after 1880. Thousands of exposed-frame reinforced-concrete factories and warehouses were erected along the Pennsylvania Railroad's northeast right-of-way in ensuing decades, and even in the year 2000, when many of these buildings lie in ruins or have been converted to discount outlets and minimalls, their thick, raw concrete skeletons dominate this extended landscape. They are by any conventional standard ugly, and they were no more generous in surface detail or variety when built one hundred years ago. Nor, when first created, were they particularly cheap compared with structures then being built by older timber-frame methods. How, then, I wondered, did these bland, hulking buildings come to achieve such a presence in our environment? To many people in business, science, and the arts after 1900, I found, they seemed the best of all possible buildings for a modernizing America. They brought science into an enterprise that many saw as economically and socially retrogressive. Where craft tradition had once held sway, technical innovation would now prevail.

No doubt the main arguments of this book violate some traditional boundaries between historical subdisciplines, but I think the end result is a sturdier fabric of historical explanation, albeit an account that cannot be easily categorized for library shelf or Web search. My disregard for disciplinarity I attribute to having been trained first in the fine arts and art history, and then, some years later, in the history of science and technology. At first my newer skills seemed to represent by far the superior field of inquiry—a chance for a more socially enlightened branch of history to "correct" older, narrower interpretations of human-made objects. But I have come to appreciate the richness that close analysis of visual form can bring to the study of knowledge systems—such as building, science, or management—and in the end I hope that I have created a mixture of the two disciplines that is more persuasive than either might be alone.

Much of the credit (or blame) for this alchemical approach must be laid on my experiences as a graduate student in the resolutely and thrillingly interdisciplinary Department of History and Sociology of Science at the University of Pennsylvania. I am deeply indebted to Thomas Hughes, Robert Kohler, Henrika Kuklick, and Judith McGaw for their conceptual and methodological guidance on this project. Their amalgamated approaches to the history of science and technology, I believe, are responsible for the richness I detected in the history of concrete and certainly for my own profound enjoyment of this discipline. Had they not directed me to additional faculty with interests in social and intellectual history, this book would have been weaker as well. Further, any felicity to the aesthetic and cultural arguments herein I owe to Peter Galison and Caroline Jones, with whom I worked with great pleasure as a postdoctoral fellow at Harvard University.

In forming the parameters of this inquiry I had much help from Bruce Seely, and I hope he considers the history of materials testing to be well served. I received vital technical advice from William Hime, Michael Scheffler, and Paul Gaudette of Wiss, Janney, Elstner Associates, Inc., and from personnel of the Portland Cement Association.

I also thank the staffs of the Hagley Museum and Library, the Baker Library of Harvard University, and the archives of Iowa State University, the University of Pennsylvania, Lehigh University, and the University of Illinois, all of whom directed me to features of this story I would not otherwise have encountered. Had the Aberthaw Construction Company not saved its corporate records and generously given me access to them, the arguments made in this book would be far less solid than they are. The materials I found there were truly inspirational.

I feel extremely fortunate to have had as close friends and colleagues throughout this project Lissa Hunt and Alex Pang, both of whom have long offered invaluable guidance on a number of levels. I also worked through countless, no doubt tiresome, features of this narrative with Dean Herrin, Emily Thompson, Abha Sur, Deborah Fitzgerald, Kathy Steen, Karen Rader, Tim Alborn, and John Zimmerman, and I thank them all.

Marissa Golden and Janet Abbate supplied crucial intellectual and logistical assistance in the later stages of this project. My sister, Deborah Slaton, and brother-in-law, Harry Hunderman, provided a remarkable combination of technical information, contacts within the field of historic architecture, and hospitality, and I owe them tremendous thanks as well.

The editors of the Johns Hopkins University Press have been very helpful, and I am appreciative of both their general support and their detailed attention to all aspects of this book. I also thank readers and commentators who have addressed portions of this narrative in the form of journal articles and conference papers over the years; I am especially grateful to Robert Friedel, Ken Alder, and John Staudenmaier for their suggestions. I also thank my colleagues in the Department of History and Politics at Drexel University for providing a congenial work environment in which to pursue this research, and Peter Groesbeck for his graphic skills.

Three people have shared the greatest burdens in the production of this book, and I want to express particular gratitude to them. First, my daughter Eleanor has provided the most wonderful company I can imagine from her appearance, midproject, five years ago. Second, my mother, Pearl Slaton, has been an unending source of editorial and emotional support. I am incredibly lucky to have had her talented, patient, and cheerful devotion to "the cause." Lastly, I lovingly thank my husband, Peter Mayes, for what can only be called heroic support of this undertaking. I could not have come close to completing this task without him: "Job done!" as he would say.

Reinforced Concrete and the Modernization
of American Building, 1900–1930

INTRODUCTION

Science and Commerce
Scenes from a Marriage

And who has to think more deeply than the engineer?
—J. A. L. Waddell, 1915

This is a book about concrete. That admission made, an even bolder one must follow: this book finds concrete to be a valuable research site in the history of American science, technology, and industrial labor. The familiar, featureless gray walls of concrete buildings, bridges, and viaducts may offer little in the way of visual pleasure, but their ubiquity reflects a complex of technical developments and social change arrayed across the twentieth century. From 1900 onward, a cadre of supremely successful technical experts—materials scientists and engineers in academic and industrial positions—devoted their careers to refining the techniques of concrete construction. With managers and owners of building firms, these experts created a series of technical protocols to carry new knowledge about concrete to the building site itself. Most important for this study, these practitioners together installed with the use of concrete a particular organization of labor in construction—one that is typical of modernizing industries in the twentieth century. As concrete replaced wood and masonry in many contexts, it brought to building an extreme division of labor that called for a relatively small coterie of highly trained specialists to supervise the work of many little-trained, and much lower-status, laborers. The case of concrete demonstrates that not only managers but scientists and engineers were instrumental in the twentieth-century modernization of productive work, an event of tremendous social consequence.

The Pursuit of Regularity

The proliferation of austere, standardized reinforced-concrete factory buildings in the United States after 1900 set stylistic and material fashions in commercial building for at least a century. Thousands of these buildings were of absolutely indistinguishable profile, alike in overall shape and surface treatment and deliberately devoid of individualizing decoration, and that uniformity is the starting point for this study. Resemblance is, of course, the very stuff of architectural history. When two skyscrapers, two mosques, or two Masai huts look alike, we recognize the assertion of an architectural style. We see the resemblances of such structures as a trail of influence and innovation occurring within a set of expressive conventions. When more prosaic structures, such as tract houses, suburban office buildings, or woodsheds look alike, aesthetic idiom seems a less pertinent index. Instead, we acknowledge a uniformity wrought by economy: to simplify and standardize is to conserve effort or expense. These two types of commonality help explain the origins of many structures and direct us towards manifestations of both creative genius and cultural change.

A third source of resemblance among buildings is science and technology—enterprises that have at their heart the goal of replicability. Technical knowledge brings to a culture new building methods and designs that come to be seen as "best practice" and that proliferate as such. In small-scale or more tradition-bound settings such a proliferation might come about simply through the passage of wood-carving or brickmaking methods from family to family or village to village. In what we might conveniently label "higher-tech" cases, scientific investigation renders entirely new materials safe and affordable, and these too bring a definitive approach—that is, one best way of building—to multiple sites. As industrialization has proceeded over the last hundred years, mass-produced construction materials, prefabricated elements, and published standards and specifications have been particularly effective in disseminating architectural "best practice" to widely disparate locales. These developments have brought a new level of uniformity to our built environments—creating local, national, and global commonalities among landscapes—and are the focus of this book.

Which of these causes of visual similarity—aesthetic, economic, or technological—we privilege in explaining any given set of formal similarities of course depends on the case. Accomplished architects have determined the form of some buildings, cost-conscious developers that of others, and manufacturers of prefabricated steel trusses and glass block that of still others. There are two important reasons for choosing for study cases of the third type, in which technology is a primary source of architectural practice. First, construction technologies have received far less attention as a source of cultural change than have aesthetic and monetary considerations. The influence of technical knowledge on architectural form has been a vital part of certain historical narratives—the development of medieval cathedrals or Bauhaus design tenets, for example. But there has been little effort to understand the role of technology in the creation of less distinguished structures. This is a particularly notable omission in accounting for modern building forms. Engineering, materials science, and industrialized construction methods have configured many buildings that did not involve primarily the talents of trained architects, or in fact any architects at all, but that fill many landscapes today with their stark, standardized profiles. Countless warehouses, factories, and other utilitarian structures of this kind appeared in the United States long before the European modernists gave functionalism their imprimatur; the indigenous cultural origins of these buildings should be recognized.

The second and arguably more important reason for looking closely at building technologies is that their historical significance goes beyond accounting for visual change. As systems of rules, modern engineering and science carry social impacts that order work and structure commercial relations. These bodies of knowledge help determine how the jobs of designing, planning, and erecting structures will be undertaken and by whom. This book is particularly concerned with the profound transformations brought to the labor of building by the rise of technical professions in the United States after 1900. For the first time, people commissioning buildings asked if the task of designing a school or factory or hospital should fall not to a self-taught artisan but to a university-trained expert. Should that expert be an architect, equipped with a comprehensive background in historical aesthetic forms, or an engineer, trained primarily in cutting-edge technical procedures? The emergence of new bodies of technical expertise had far-reaching effects for the building trades as further questions of occupational authority took shape. Once designed, shall buildings be constructed by highly skilled masons, carpenters, and glaziers using traditional methods and materials, or by less experienced laborers pouring concrete and assembling prefabricated trusses or windows? New technologies brought a new understanding of what knowledge counted as authoritative on the building site, and what as retrograde and unreliable.

Proponents of concrete answered such labor questions with a resounding vote for a modern, hierarchical organization of building work. With few exceptions, concrete buildings were erected with large numbers of little-trained workers operating under the supervision of a few highly trained, and well-paid, employees. The increasing role of science-based testing and inspection in concrete construction between 1900 and 1930 is reflected in the significant managerial presence of those trained in these technical specialties. Existing systems of hiring, supervision, and training, based in old family and regional networks within the building trades and sustained by craft unions such as those of bricklayers and stonecutters, had little influence in the concrete industry, and technological advance helped render their diminished role both possible and permanent.

To comprehend this alteration of modern work, we must first recognize that between 1890 and 1920 the idea of uniformity came to exert a remarkable hold on American industrialists. The era brought an unprecedented level of regularity to the quality of raw materials used in industry, the procedures of production, and the character of finished products. In manufacturing and food processing, in mining and construction, predictability—and, not coincidentally, profits—reached heights not seen before. This transformation can be pictured in part as a physical one. Tidy assembly lines replaced cluttered workbenches; standardized screws and teakettles replaced idiosyncratic models; the steady flow of wet concrete replaced the piecemeal accumulation of bricks. But it was indisputably also a social transformation, in which the work experiences of many Americans changed dramatically. For the highly paid technical and managerial experts of the period, as well as the much lower-paid rank and file, the daily tasks of industrial employment called for new skills and workplace behaviors. More broadly, firm hierarchical structures of occupational authority took root and configured long-term patterns of social and economic mobility. Finally, this pursuit of order can be seen as an intellectual or epistemological development, grounded in a wide embrace of technical or scientific knowledge for commercial purposes. Because the industrial commitment to science provided so many mechanisms of change in the experiences of American workers, it is important to explain the social instrumentalities of industrial science.

To see the growth of American industry after 1900 as a pursuit of regularity—rather than of scale, scope, or efficiency per se—is to sense the immense organizational power brought by scientific thinking within commercial contexts. This was true not only of the research and development work that brought a startling new inventive capacity to American industry after 1900 but also of a much less celebrated but more common commercial use of science: the routine application of scientific and technical knowledge to production. Sometimes labeled by its promoters the science of materials, sometimes claimed as a branch of engineering—from 1900 to today the meanings of these terms have continued to shift—such applications can be grouped under the term *quality control.* This emerging field included a wide range of testing and inspection practices, the operations of gauging parts and grading materials, and the creation and use of written instruments, such as standards and specifications, that disseminated these technical protocols to shop floors, mines, refineries, and construction sites. Because quality control was so pervasive in industrial production, it was in many ways more influential than research and development in spreading an ideology of order and precision. This was science working to enhance the productivity of industry; it was also industry offering unprecedented opportunities to scientific and technical occupations.[1]

In retrospect we know this to have been a fruitful pairing. Science-based quality-control techniques pervade modern industrial practice. The sturdiness of buildings and the purity of foodstuffs derive from such measures, as do the interchangeability of automobile parts and the predictable fit of railroad cars to railroad tracks. The issuance of industrial standards and, often, their enforcement were carried out through specialized private and public scientific bodies that continue to proliferate today. Yet this long, happy marriage of science and commerce was by no means assured from the beginning. It came about despite some apparently profound contradictions between the ideologies of the two enterprises. Historians have addressed the sometimes antagonistic mutual influence of science and commerce in the upper reaches of industrial research and development, where conflicting values about secrecy and profit faced resolution.[2] The history of routine science-based quality control reveals not only further tensions and accommodations between these enterprises but also engines of the broad transformations of work described above.[3]

Various such negotiations between science and commerce reveal the social dimensions of industrial quality control with particular clarity. To begin with, a paradox surrounded the preservation of intellectual authority in the world of industrial employment. In the expanding post-1900 economy, select bodies of knowledge retained an elite character even as they achieved a true commodification, moving outward from their academic origins to wide application in the world of commercial production. To a degree, chemistry, electrical engineering, and certain social science disciplines achieved such a market identity, but the scientific understanding of materials and machines carried this pattern to an extreme, attaining a utility for industry on the scale of financial or managerial expertise. Such knowledge penetrated commerce in two ways. In some instances, university faculty undertook the testing of materials, incorporating specimens of commercial products into classroom work or performing tests as paid consultants on their own time. Far more frequently, young men trained as testers and inspectors moved from university engineering programs into salaried and often supervisory industrial positions. In this dissemination process certain features of quality-control work assumed a "rationalized" form, as testing kits, simplified instruments, and published materials standards multiplied. Yet dedicated testing and inspection positions—distinct from the rudimentary inspection tasks assigned to line workers or foremen—remained almost exclusively a domain of college-trained men. No matter how great the market for applied knowledge about industrial processes and materials became, the actual applications of such techniques did not pass downward into the hands of rank-and-file workers. A college-trained specialist in concrete testing in 1908 received a salary two to four times greater than that of a foreman on a construction site, and up to eight times that of an ordinary laborer.[4] Yet despite this economic incentive, even simplified concrete-testing procedures did not become the purview of foremen or laborers. Rather than regarding such divisions as inevitable, or as divorced from the design of technical expertise, we may ask by what means scientific testing and inspection evaded the extreme rationalization and economic devaluation common to so many forms of industrial labor.[5]

In part, quality-control specialists on the modern work site benefited from the aura of scientific authority created by university-based research and training programs. The commercial success of the degree-holding technicians reflected a certain glory back on the academic scientists who had trained them in a self-perpetuating system of boosterism for the new specialty. But we still need to ask why the proffered public image convinced its audience, the industrialists who employed modern quality-control specialists. Finding an answer to this question will do more than explain the good fortune of this single scientific field, for the implementation of new quality-control methods carried different economic implications for different groups of working people. Technical skills and opportunities in the modern construction industry often divided along lines of gender and ethnicity. These divisions suggest that the ascendant status of materials experts in industrial ranks dovetailed with social agendas of broader consequence—conceivably agendas shared by the experts and those industrialists who willingly hired them despite their relatively high cost.

The form and content of scientific knowledge prepared for use in commercial settings asserted this arrangement of social privilege. Relative to the narrow, repetitive character of most industrial work, that of testers and inspectors comprised knowledge both comprehensive and cumulative. Somewhat ironically, science-based quality control resembled in these structural features the integrated work done by traditionally skilled artisans in older productive settings. Like corporate managers of the era, materials experts claimed a broad knowledge base for their work, even as lower-echelon jobs were being famously divided into small, repeated operations of very limited scope. Unlike their artisanal forebears, the testers and inspectors trafficked in nontraditional bodies of knowledge such as scientific theories and protocols, modern instruments, and data on the behavior of materials. But professions, including technical professions, working for industry in this period often celebrated their integrative, analytical capacities as an almost craftlike expertise.[6] Such an appeal to both quantified and unquantifiable techniques seems, at first, to reflect opposing professional strategies on the part of the materials experts. The one strategy accommodated the economizing impulses of their industrial clientele; the other, seemingly retrogressive, asked industry to pay for something it could not measure. In fact, the two approaches were complementary, giving materials experts both wide appeal and unique value in the competitive commercial sphere.[7] While social critics such as Thorstein Veblen celebrated a realm of academic inquiry free from the venality of capitalist competition, these experts busily melded the rewards of both domains.[8]

Their retention of unquantifiable techniques aided materials experts in their encounter with a second paradoxical condition of industrial science: the problem of scientific certitude in a world of shifting competitive advantage. In part this was a matter of maintaining the certitude traditionally granted to scientific investigation—of imparting an objective character to scientific practice as it served for-profit enterprise. To materials experts, testing technicians, and those who employed them, matters of neutrality and trust were understandably of great import. However, *limiting* the essential fixity of scientific findings was also extremely important to competitive industrial operations. The efficacy of standards, instruments, and technical protocols, to say nothing of the actual findings of tests and experiments, is predicated on the idea that science brings into being a fixed body of knowledge. But production enterprises, and those supplying services to such businesses in a competitive environment, must function in the realm of reputation, where the immutability of fact can be a disability. In the case of quality-control operations, negative test results or the emergence of new products and new rivals might prompt a business to remake its public identity. Firm data about materials, products, or the performance of services would seem to prohibit such refashioning. For materials specialists working for industry the question became how to import science into production while allowing for an infinitely changeable definition of "best practice."

To contend with this tension, materials experts again exploited qualitative features of their practice. In pedagogical and professional settings they emphasized the significance of "character"—the high moral caliber, the sheer trustworthiness of the quality-control man—as a foundation of their services. In their design of engineering curricula and of protocols for industrial application, the experts instantiated the notion that it was *the tester, not the test,* that assured quality production. This construct not only tightened their control of the occupation of testing but also created a technical practice amenable to the needs of business. The scientific quality control of industrial operations was, from its inception, an inherently subjective project.

While the competitive conditions of commerce offer one level of explanation for this subjectivity within technical practice, here, as in the establishment of an elite status for this technical specialty, the shared social visions of experts and their industrial clientele must be acknowledged. Had industrialists not started with some conception of who might best do what task—male or female employee, young or old, black or white, German or Polish—the idea of practitioner overriding practice would not have made sense. The central question may be why, and indeed whether, Americans believed that the presence of scientific rigor guaranteed political or social neutrality in this period. (And why has this ascription of neutrality been so rarely problematized by historians?) What exactly did *neutrality* mean in this context? As David Noble and others have indicated, leaders in the technical professions of the new century envisioned disparities in the hiring, promotion, and productive capacities of different genders and ethnicities. Many such visions were enacted.[9] Certainly laboring trades themselves functioned with pronounced gender and ethnic divisions in this period, and we should not treat the biases of industrialists as some exogenous factor in American society. But by studying closely the emergence of science-based quality-control methods we can locate at least some of the mechanisms by which, for many people after 1900, extreme divisions of occupational opportunity came to seem like natural features of a prosperous industrial economy.[10]

Concrete and Control

This book explores the social origins and consequences of industrial quality control through the case of commercial concrete construction between 1900 and 1930. This was a field of self-conscious and rapid modernization. Longstanding artisanal technologies such as bricklaying and carpentry gave way in many instances to the large-scale and strategically managed techniques of prefabricated and reinforced-concrete construction. The use of reinforced concrete in particular grew as a commercially viable technology through the efforts of scientific occupations. From an intriguing but cumbersome and expensive material in the 1880s and 1890s, concrete became by 1920 by far the favored medium for large building projects, especially for the massive utilitarian buildings required by the flourishing manufacturing sector. Concrete emerged as a subject of intensive science-based testing and inspection around 1900, with huge numbers of technical standards and specifications being created to direct those investigative tasks and the actual handling of this scientifically controlled material on the modern construction site. In each regard reinforced concrete reflected its purveyors' conceptions of how productive labor would best be organized. Even the architectural forms assumed by concrete in the starkly functionalist factory buildings that filled American industrial neighborhoods after 1900 were a blunt expression of industrialists' embrace of scientific information and expertise, and of the rising cultural status of both science and industry.

The role of science in the transformation of modern productive labor is understudied. Historians have considered the rare, cutting-edge advances that research scientists brought to industry in this period, but not the daily routines of scientific testing and inspection exemplified in the newly efficacious use of concrete. I follow here concrete's inception as a set of laboratory problems, its institution as a subject of university curricula, and its ultimate instantiation in a series of workplace initiatives for "quality control." In so doing, I carry to the everyday technical work of production questions already asked by others of experimental science: How do the institutional conditions under which scientists or engineers practice—the laboratory, the corporation, the foundation—shape scientific inquiry? How do scientists shape the institutions in which they work? Further, how do the materialities of intellectual inquiry—such as instruments and the physical features of materials under scrutiny—shape that inquiry, and vice versa? Perhaps most intriguing is a set of questions only recently explored in the history and sociology of science and technology: How do epistemological trends—such as increasing precision, quantification, or standardization—embody the social visions to which scientists and engineers (and their employers) subscribe?

Throughout, this book approaches the increased use of reinforced concrete as the product of intertwined intellectual and social developments, linking the technical refinement and application of this new material to the occupational agendas of its promoters. Chapter 1 traces the development of testing and inspection methods for concrete by materials scientists in American public and private universities. These experts taught succeeding generations of practitioners. They also created a body of knowledge about concrete—tests and testing machines, criteria for performance, written instruments for the control of concrete's quality on the construction site—intended for a growing audience of commercial engineering and building firms. Crucially, the materials scientists, working simultaneously as instructors and within professional societies, wedded the industrial use of scientific techniques to particular practitioners. They designed university curricula for prospective testers and inspectors, and the content of technical standards themselves, to associate materials testing only with college-trained technicians. Their efforts were successful. University graduates would dominate the field of concrete testing and inspection in the new century.

Essential to the scientists' success was the dual nature of the curricula and materials standards, both of which combined the use of highly routinized and quantified procedures with invocations of subjective, experience-based judgment. Chapter 2 follows this development. In the classrooms and laboratories of university engineering programs in materials after 1900, instructors sought to endow their students with a blend of rote technique and what we must see as the largely inchoate properties of intuition and powers of observation. Academics bolstered these pedagogical designs with expansive claims that engineering graduates possessed an elevated character and deserved appointment as cultural stewards in a modern "technological" society. No accolade was too lofty for these "special citizens," soon to "direct the great sources of power in nature, for the use and convenience of man."[11] Their intellectual prowess was the subject of constant celebration; these were deep thinkers indeed. Enrobing testers' technical capacities—again: routine enough to be affordable to industry—in such a heightened cultural status helped defend the discipline from incursions from below, since lesser-trained people conceivably might have done such work. Negative depictions of other groups of industrial workers aided the task of gatekeeping. Instructors deemed certain ideas of manliness (which helped exclude women from careers in quality control) and "inherited" characteristics (meaning race or ethnicity) to be requisites for good engineering. These ideas take up nearly as much space in pedagogical discussions of the period as descriptions of instructional equipment and exercises. The consulting work of academic concrete experts similarly reflected a sensitivity to the wider "moral economy" of industrial science, here expressed as constant movement between cooperative and proprietary activities and other strategies of a doubled self-definition.[12]

The teaching and consulting work of concrete experts demonstrates that by including calls for experienced judgments, bodies of scientific knowledge about concrete could establish relatively narrow qualifications for their own use. This exclusionary function for materials science extended to the standards and specifications authored by academics with their industrial and governmental colleagues at the American Society for Testing Materials, the American Society of Civil Engineers, and similar smaller associations.[13] Chapter 3 provides a study of the content of standards and specifications for concrete between 1900 and 1930. The frequent omission of absolute values for concrete specimens and the rejection of mechanized testing and handling methods in these protocols echo the definition of testing skills laid out in the university. In many regulations recourse to remarkably subjective language reiterated the testers' discretionary abilities. The lack of precision in these written instruments was not a developmental problem, any more than the flowery claims of the engineering instructors simply reflect linguistic conventions of the day. Rather, the creation of standards brought about tremendous economies and new levels of safety in building while institutionalizing very particular allocations of scientific and social authority.

Standards and specifications held additional utility for the architectural trades as a tool for the control of interfirm competition and public image.

Written protocols for technical work, embodied in contracts, estimates, and even advertisements, carried science-based knowledge into the realm of liability and business reputations. Building designers, the engineering firms that usually supervised concrete building projects, and the contractors engaged to erect concrete structures functioned in an arena of high-stakes credit and blame, and all relied heavily on specifications to navigate this world. Among the protean features of science-based testing protocols that emerge here are shifting definitions of what constitutes a reliable test procedure or result; conflicting ideas about whether payment for testing services guarantees or corrupts scientific objectivity; and debate over which test findings shall reach the general public and which shall remain proprietary, and thus controlled, records of a commercial concern. Here the issue of scientific fixity receives close attention. The definitive nature of science-based testing for cement and concrete products could saddle producers with a burdensome negative reputation. Objectivity, in the senses of both fairness and precision, became a constant concern and a term of constantly shifting meaning. Without question the scientific imprint of testing and inspection held great power in the world of commercial construction as marketing tool and competitive weapon.

To explain fully the construction industry's acceptance of such a high valuation of science, we must move from the consideration of standards and specifications to the daily operations of construction: the material and organizational challenges faced by companies that built with concrete after 1900. In chapter 4, which examines the administrative and technological choices made by owners of large factory-building firms between 1900 and 1930, we learn why it was that commercial builders, presumably bent on reducing wages wherever possible, favored the association of quality control for concrete—in most other ways a material subject to "true" mass-production handling—with an elite and relatively highly paid cadre of testers and inspectors.

One explanation that emerges from this "ground-level" focus on concrete construction is that the competitive conditions of commerce also embodied a large degree of social consensus. Both materials scientists and the construction firm operators who used their findings and employed their students brought to their work a hierarchical conception of modern technical skill. University engineering curricula, technical literature on concrete, and the administrative policies of building firms all expressed their authors' belief in the innately unequal technical capacities of men and women, of white and nonwhite workers, and of native- and foreign-born workers. White, native-born, and for the most part Protestant males were widely seen to carry superior potential as technical practitioners. Standards and specifications that made up "codes of best practice" for reinforced-concrete construction embedded these perceived differences in the daily activities of the work site. For example, many of the protocols explicitly deemed the activities of testing and inspection worthless if undertaken by the "wrong type" of worker. Thus, what at first glance appear to be socially neutral tools of enhanced efficiency on the construction site actually promoted specific delegations of responsibility, credit, and mobility for different factions of the workforce. In protecting their own enhanced status in the world of commercial construction, materials experts supported the social visions of their clientele as well.

These social visions found cultural expression and reification in the outward form of reinforced-concrete buildings erected for American industries in the new century (fig. I.1). Chapter 5 considers the appearance of these highly standardized buildings, investigating their builders' and owners' rejection of traditional architectural forms and embrace of a modern aesthetic alternative. Concrete factory buildings conveyed a mixture of progressive and conservative cultural messages. On the one hand, their builders rejected historicizing "high-culture" architectural reference for the visual celebration of modern materials and construction techniques. The austere reinforced-concrete structures were both the products and the overt expressions of the efficient, forward-looking techniques of modern mass production. Much to their owners' pleasure, the stark, standardized factory exteriors echoed the streamlined productive work going on within. But the labor of both factory construction and the manufacturing occurring within the factories carried social consequences of a distinctly unprogressive character. The functionalist styling of these buildings actually celebrated a new hierarchy of cultural achievement. The work of creating standards and specifications—of standardizing techniques or designs—was itself a privileged expertise, not subject to the diminished autonomy or mobility of most industrial work. So too the overall use of modern quality-control methods and the general operation of modern industrial concerns reasserted such social divisions. In their scientific genesis and their appearance, reinforced-concrete factory buildings reflect the complex of social consequences that accompanied science's entry into the world of routine—but hardly ordinary—industrial production.

FIGURE I.1. Reinforced-concrete factory complex erected by the Aberthaw Company for the Samuel Cabot Company (makers of creosote and related chemical products), Boston, ca. 1909. Exposed-concrete structures of such functionalist appearance, undisguised by traditional brick cladding or other decorative features, rapidly came to dominate American industrial architecture in the decade following 1900. *ACC Archives*

This is an approach to understanding the built environment that is unlike much existing work in the field of architectural history. It introduces not only a strong technical component but also a powerful set of social forces into the time line of aesthetic modernism. This emphasis on the experiences of workers, at all levels of training and occupational authority, seeks to enrich our perception of how buildings come to be. More important, perhaps it can illuminate the powerful mutual influence of scientific practice and social organization, combining traditional labor- and social-history concerns with methods of some newer history and sociology of science. As resistant to exploration and complexity as concrete may seem as we daily drive by and over it, for the inquiring historian it reflects a multitude of social and cultural forces.

A Brief History of Concrete

Some technical information may be helpful in following the history of the reinforced-concrete factory building. The following pages define terms that are used throughout this book, and provide a brief overview of concrete building prior to 1900. The development of concrete reflected a long history of technical experimentation and, at the end of the nineteenth century, a remarkable burst of entrepreneurial enthusiasm that brought it into wide use among American builders.

Cements are powders comprising naturally occurring rock deposits (so-called natural cements) or combinations of ingredients (artificial or "Portland" cements), all of which include clay and lime in varying proportions.[14] In the preparation of Portland cement, raw ingredients are ground to a powder, burned in kilns, and then milled. When mixed with water and allowed to harden, cement creates a material of great compressive strength and, depending on its chemical composition, one that may also harden under water. *Mortars* are combinations of sand, cement, and water used to bond stones or bricks. *Concrete* consists of gravel (or, less commonly, crushed shells or cinders), sand, cement, and water blended to create a liquid medium that can be poured into place, where it then hardens, or cures, to achieve a solid mass of great compressive, but little tensile, strength. Iron or steel reinforcing bars embedded in concrete while it is wet can provide tensile strength for the hardened mass.

The use of natural cement for the production of mortars was first known among the Romans, disappearing during the Middle Ages to be rediscovered in England in the mid-eighteenth century. The use of cement and concrete in the United States began with the cement mortars applied to the construction of canals, tunnels, and bridge abutments in the second quarter of the nineteenth century. This was followed by the application of cement to buildings in the form of precast concrete blocks, which could imitate traditional masonry in function and appearance. These were first used around 1840 and were mass-produced after 1868.[15]

The considerable compressive strength displayed by concrete in cast-block or poured form inspired mid-nineteenth-century engineers to employ it for piers, walls, footings, and paving. By the 1880s, American engineers were employing Portland cement concrete for piers and abutments of very large bridges and for compressive members of multistory buildings. At the same time, concrete was found to be a valuable material for fireproofing structural steel, since it was less expensive than established methods of ceramic cladding. In this application, steel and concrete worked well together, adhering to one another dependably over time. Builders' confidence in the physical compatibility of steel and concrete prepared the way for their acceptance of concrete reinforced with metal bars, an addition that gave concrete the ability to withstand the bending, or tensile, forces to which floors, ceilings, beams, and columns are subjected within a building. In the first decade of the twentieth century, reinforcing technologies greatly expanded the use of concrete in the design of American buildings.[16]

Reinforcing had its own lengthy history by the time it was enlisted for concrete. Clay, stucco, and other earthen building mediums had been strengthened by the inclusion of straw since antiquity, the straw helping to bind the friable materials. The earliest attempts to reinforce masonry beams with iron were made by French architects in the late eighteenth century, followed by English and American efforts in the early nineteenth century. Engineers and architects in all three countries began to establish an empirical understanding of the behaviors of iron hooping and meshes in concrete of various forms.[17] Americans first received patents for reinforced-concrete piping and timber-reinforced walls in the late 1860s. In 1875 the mechanical engineer William Ward built a house of concrete beams reinforced with iron rods and I-beams in Port Chester, New York, cited as the first successful attempt to craft an entire building of reinforced concrete.[18]

Significantly, Ward placed the reinforcement at the bottom of beams, where concrete was least able to absorb tensile stresses on its own. He did not, however, translate his successes into commercial enterprise. The first efforts in this direction were those of the French engineer François Hennebique, who developed a system for rapidly producing reinforced-concrete floor slabs in the 1870s. In 1892 Hennebique patented a process of bending reinforcement bars to better resist the tension that occurred in concrete where it rested on supports. He soon closed his Paris construction firm in order to open a consulting firm that trained and licensed contractors in his innovative technique.[19]

Hennebique's innovations in the design and marketing of concrete buildings had a significant impact on American practice, particularly in the development of systems of prefabricated reinforcing elements and their sale by licensed contractors.[20] It is important to note, however, that work on the design of concrete-frame buildings was occurring in the United States throughout the last third of the nineteenth century. In 1877 the American engineer Thaddeus Hyatt described the principle that reinforced-concrete had to resist enough tensile stresses to balance existing compressive stresses. His research led to enhanced understanding and control of reinforcement in slabs, beams, and columns.[21]

Historians laud Hyatt as an inventor and a theorist but give credit to builders for the practical development of reinforced concrete in America.[22] Among the most frequently cited builders is Ernest Ransome, who began his career in California after 1870, addressing the need for earthquake-proof buildings. In 1884 Ransome patented a type of twisted rod that was both inexpensive to produce and extremely effective as reinforcement. During the 1880s and 1890s he focused on creating floors that did not require support from iron or steel beams or columns, relying instead on combinations of concrete beams and slabs and on systems of concrete ribbing.[23]

Like Hennebique, Ransome was a dedicated entrepreneur. Throughout the 1890s he marketed his services as a consultant and licensor, creating a successful business that disseminated his technical knowledge and inspired related innovations by competing engineering firms. By the turn of the century, dozens of companies were marketing systems for reinforced-concrete construction that featured various arrangements of mass-produced metal rods for the reinforcement of beams; netting or fabric for wall, floor, or roof slabs; and hoops and spirals for columns. Among the best known were systems developed by Albert and Julius Kahn and C. A. P. Turner, but many others were designed and aggressively marketed (these are described in chapter 4).[24]

Ransome described his own work of the 1890s as marking a transition in the design of industrial buildings, "the closing of the old-time construction of reinforced-concrete buildings, constructed more or less in imitation of brick or stone buildings, with comparatively small windows set in walls."[25] New factories and warehouses that exploited the carrying capacities of concrete frames offered, instead, large expanses of glass. Certainly Ransome's role in this stylistic shift to the "daylight" factory was substantial, but he did not work in isolation. The architectural conception of concrete in America was shifting away from the imitation of masonry for a number of reasons. Some of these reasons had to do with technical developments, such as the popularity of concrete for fireproofing, while others had to do with the changing nature of the design professions. In the 1890s, concrete-block construction lost the appeal it had initially held for architects of residences and civic buildings. The mass-produced blocks came to be associated with amateur building design, especially after the material was made available to the general public through mail-ordered concrete-block machines that could be used by the untrained.[26] At the same time, concrete's potential as a structural element was drawing the attention of designers of utilitarian buildings through its integration with existing structural steel design (the popularity of steel-framed factories and warehouses after 1880 is discussed in chapter 4). In the early 1890s, when inexpensive systems of expanded metal and mesh concrete reinforcement came onto the market, architects and engineers could incorporate concrete into their steel structures as the filling and arch work between beams, replacing the more expensive brickwork or terra cotta traditionally used.

Through this relatively conservative introduction, designers of heavy-duty buildings became familiar with the structural properties of concrete. American cement manufacturers exploited and encouraged this new interest, and the annual domestic production of Portland cement grew from 300,000 barrels in 1890 to 900,000 barrels in 1895.[27] In 1896 the American production of Portland cement surpassed 1 million barrels for the first time; by 1906 it had reached 46 million barrels per year. In the same period, the number of commercial cement plants in the United States grew from twenty-six to more than one hundred, representing invested capital of more than $100 million. This was not a temporary growth spurt: in 1924, American cement companies shipped 146 million barrels of cement.[28]

A combination of technical and commercial changes contributed to this growth. The efforts of American cement manufacturers to compete with European firms brought American builders a product of higher quality and lower price than they had known before, and supply and demand both rose steadily. Presented with the decreasing cost of cement and related materials, engineers found it worthwhile to focus on the new technology. The resulting advances in methods of reinforcement permitted the use of concrete in larger and more complex structures than had previously been possible.

New systems for the preparation and handling of concrete helped as well. Traditional methods of producing cements had included grinding and milling by water power, air-drying slurry on heated floors, and burning the resulting powder in dome kilns. In the 1890s, American manufacturers introduced kilns (developed in England) that not only could be operated continuously but also combined operations of burning and drying with grinding. This emphasis on flow was embedded in new methods for continuous mixing and delivery of wet concrete around construction sites, and large users of cement and concrete, particularly railroads and federal and state governments, executed huge projects such as the Panama Canal (completed in 1915), countless railroad bridges and gradings, and miles of paved highways. The material's popularity encouraged cement producers to invest in new plants and commercial distribution systems, rendering concrete still cheaper for building and engineering projects of all sizes.[29]

With the commercial embrace of concrete proceeding rapidly, an incipient professional movement among engineers and materials scientists to address technical features of the medium also expanded. This movement in turn encouraged further production and use of cement and concrete materials, and the stage was set for a broad modernization of commercial building in America, with the organization of building labor and architectural styling changing alongside the materials and methods of construction.

CHAPTER ONE

Concrete Testing
The Academics at Work

One is very near to nature's heart when making tests.
—Charles Dudley, 1906

There is a persistent tendency among educational analysts and historians to divide early-twentieth-century engineering fields into those based on conventional shop procedures—the so-called cut-and-try approaches to engineering—and those that developed more "scientific" methods. William Wickenden's widely read 1930 study of engineering education, for example, associated new fields such as electrical and chemical engineering, founded in the 1880s and 1890s, respectively, with a scientific approach. By this he meant that they used a consistent theoretical framework, which he attributed to the fields' leaders having been trained in physics and chemistry. In contrast, Wickenden saw civil and mechanical engineering as having a less theoretical emphasis.[1] Edwin Layton, in his more recent overview of American engineering, maintains that hydraulics and the study of the strength of materials after 1900 both "built directly on science" (and were thus appropriately classed by some educators as branches of physics).[2] Each observer touches on important ways in which engineers defined their expertise in this period. However, the dichotomizing pictures these narratives create of engineering in the new century are misleading. They associate any surviving subjectivity in engineering disciplines with a retrograde sensibility or failure to achieve a desired modernity, and they suggest that "scientizing disciplines" rejected all attributes of the old "rule-of-thumb" practice, which was not the case.[3]

Many engineering fields did lose much of their trial-and-error character as they expanded their presence in the academic and commercial domains. However, the use of scientific methods such as consistent theoretical framework, systematic experimentation, and instrumentation by a discipline did not displace every existing element of traditional engineering practice. This becomes abundantly clear when we examine the educational process by which older generations of engineers trained new ones. Our focus here is on the study of materials, an area in which research on concrete found a home after 1900. In this field, and perhaps in others geared towards industrial application, training juxtaposed solidly systematized and minimally defined operations. Among university engineering instructors, mathematical formulas, theories, practical experimentation, instruments, and laboratory reports were all invoked to some degree as appropriate means of conveying skills to students. At the same time, however, certain judgments were considered to be matters of "experience" and "intuition" alone, essentially unteachable and arguably unscientific, if "science" is taken to mean only systematized or predetermined practices. "Learning by doing" maintained its viability as a component of modern university instruction despite its resonance with longstanding rule-of-thumb techniques. While working on new materials and industrial problems, the concrete experts behaved, in selected circumstances, as technical thinkers of earlier generations had done. They held to longstanding conceptions of what made a qualified practitioner.[4]

The retention of ad hoc, or "prescientific," methods in engineering education was not accidental; it should not be mistaken for ignorance on the part of educators about what activities constituted precise or effective teaching or research. It was, instead, an appropriate strategy for technical experts seeking to assure their position, and that of their students, in the world of industrial employment. Educators' emphasis on good judgment, discretion, and intuition helped solidify the positive reputation of engineering graduates among their employers. Engineering instructors used that reputation to build a superstructure of moral and cultural superiority over other groups of industrial workers, and all of these claims found expression not merely in the rhetoric but in the technical practices of engineering education.

Study of the strength of materials, emerging in universities after 1900 as a burgeoning subfield of engineering, provides an exemplary case of this Progressive Era occupational self-fashioning. Industrial interest in the control of raw materials and finished products had increased steadily since the mid-nineteenth century as production grew in pace and scale. Academic commitment to serving the needs of industry expanded in the same period, so that both the demand and supply sides of "materials science," as it was eventually called, flourished. A single material, concrete, offers further focus for this narrative. Use of it as a viable material for large-scale construction exploded after 1890, and university materials departments hastened to include it among their objects of study. With steel, brick, machine metals, and a wide selection of chemical products, concrete and its active ingredient, cement, captured the attention and resources of university engineering departments dedicated to the advancement of commercial materials. In some areas, such as upstate New York and western Pennsylvania, where high-quality components of cement occurred naturally, the cement industry had a foothold by 1900, and local engineering departments found a ready audience for research and testing services. In many parts of the Northeast, the Midwest, and California, concrete was by this time finding a growing market among builders, and any locale in which commercial construction or civil engineering projects thrived might present a clientele for a university cement-testing laboratory. The largest materials investigation facilities, such as those at the University of Illinois and the University of Pennsylvania, drew on a national clientele, providing vital information to an extensive audience of cement producers, construction firms, and public works departments.[5]

In the materials laboratories of American universities, instructors undertook three tasks (discussed, in turn, in this and the next two chapters). First, with other academic specialists in the industrial application of materials such as steel, machine metals, and ceramics, the concrete specialists trained succeeding generations of scientists and engineers—the degree-holding practitioners who would take techniques of testing and inspection into positions of industrial employment. Second, university instructors helped write the published standards and specifications for concrete construction that accompanied the young testing engineers out into commercial building sites, closely directing the application of science by the building industry. A third type of activity undertaken by university engineering faculties involved consulting and advising to the building industry through research, specification writing, and promotional work. In all three functions, academic scientists helped construct a definition of quality control that preserved its craftlike character while establishing its wide utility for industry.

In training young men for jobs as concrete testers and inspectors, and in their other work for concrete trades, university materials instructors confronted the conditions under which American businesses functioned in

1900. In part these conditions centered on preventing waste of materials and labor—the problem of achieving the greatest economies that safety would allow (*safety* here referring to that of capital investments in buildings, machinery, and goods as well as that of people). But all such economic concerns unfolded in the broader context of work relations. As had occurred in many other industries, owners and managers of construction firms had, by 1900, embraced modern methods of work organization. These methods involved mechanization, and the division of building tasks as far as possible into two categories: those deemed to require a great deal of training and a comprehensive body of knowledge for execution, and those seen to entail only the repetition of very simple tasks. The first category included all management and design tasks; the second, most jobs of materials handling or assembly. Some older types of building skills, such as those of masons, welders, and plumbers, persisted to a degree, but because lesser-skilled jobs commanded lower wages, the majority of man-hours expended on building sites shifted to the second category. Materials experts sought to secure their place, and that of the young men they trained, in the first category, where they would be assured of ongoing employment, good salaries, and the social status of white-collar employment. All but the smallest building firms were enthusiastically pursuing new systems of highly rationalized work and materials, but experts in materials testing and inspecting envisioned for their occupation an exemption from such deskilling and categorization, instead, as purveyors of supervisory services.

At first glance, an elevated occupational status for testers and inspectors seems like a natural state of affairs; these practitioners worked with complex bodies of scientific knowledge using skills they had acquired through a lengthy and rigorous education. But we must problematize that characterization on a number of levels. Close scrutiny finds that many features of concrete testing and inspection were in fact "rationalized" from their inception—that is, mechanized or otherwise reduced to routines that could be easily repeated in the manner of the simplest manual labor. After 1900 many tasks of materials testing for manufacturing and construction moved rapidly from university or industrial laboratories into daily use in the field, either factory or construction site, in the form of kits, instruments, and simplified technical protocols. Why did not more testing operations follow this trajectory in order to render the entire operation suitable for minimally trained employees? How did concrete testing remain firmly in the realm of expert knowledge, receiving the exemption its promoters sought?

University instructors managed to associate with science-based testing and inspection one operational feature that by definition could not be rationalized: the tester's use of intuition and other highly subjective mental faculties. With this gesture engineering educators began the task of keeping such work in the hands of specially qualified personnel, saving it from the virtually indiscriminate assignment to which fully routinized industrial tasks were subject. Materials experts severely narrowed the population eligible for such qualifications by claiming that such valued abilities could be acquired only through exposure to university engineering curricula.[6] Coursework incorporated training in up-to-date methods of testing and inspection, including the use of new instruments and the testing of actual industrial specimens for paying customers. The ability to follow technical regimens was a vital component of university engineering programs. However, in describing their curricula and designing their courses and laboratory exercises, university materials instructors added a second type of criterion for testers and inspectors: a facility in rendering subjective judgment. In the eyes of industrial and government employers seeking reliable personnel for testing and inspection positions, only university training inculcated the needed array of "mental habits."[7]

Engineering instructors were self-consciously creating a commercial jurisdiction, establishing a particular identity in the eyes of nonspecialists who would then, it was hoped, hire university graduates and define testing as work deserving high pay and prestige. To follow their successful program of self-definition and persuasion, we need to understand what constituted good and bad engineering for these academic experts. Familiar associations of "old-fashioned" and "modern" practice with rule-of-thumb and scientific techniques, respectively, become obviously problematic here. To be "practical," for example, an instructional technique had to prepare a student for industrial employment. Exercises that replicated the physical conditions of industry, or that put students to solving actual problems posed by industry, fulfilled this function. But so too in this context did exercises that enhanced a student's "character" through some inchoate process of moral uplift. It becomes crucial that we not automatically associate practicality with quantitative operations. Nor can quantitative methods be associated with precision or reliability, since the operator's subjective capacities might be needed to assure either of those. "Hands-on" and theoretical work similarly held shifting status in the world of engineering pedagogy.

Without diminishing the technical advances brought to construction by university programs for concrete testing, we will tie these epistemological shifts to their social purposes and consequences. Engineering instructors had an occupational vision that carried with it societal implications. Since university engineering programs were at this time largely open only to men, and generally to native-born white men who had the economic means to attend college, this construction of concrete testing tied the possibility of employment as an industrial quality-control technician to one's identity as a middle-class white male. To follow the new scientific quality-control procedures faithfully, manufacturers had to hire people of a particular academic background, and thus of a particular social background. In effect, the university served as "gatekeeper" for advanced technical occupations, and in the process of shaping scientific quality-control methods, technical experts embraced and reiterated the social patterns they found in the university. We will later see employers' acceptance of this stratification—their willingness to pay more for testing services—as a ratification of social divisions. The producers (academic experts) and consumers (the building industry) of techniques for concrete testing shared a body of social values enacted by those techniques—in the hands of young testing engineers—in the workplace. We begin with a close look at the "supply side" of these science-based operations: the academic engineering programs of 1900 to 1930.

Universities Serving Industry: Institutional Origins of Materials Testing

The notion of linking instruction to the needs of industry in America received its first systematic applications in the mechanics institutes of the 1820s. The Franklin Institute in Philadelphia and similar educational institutions operated on the premise that methods of scientific study could be taught to artisans and mechanics in such a way as to benefit invention and manufacturing in this country. The only other organized dissemination of engineering knowledge at this time occurred as people involved with the construction of the Erie Canal went on to other civil engineering projects, both public (such as canals or roadways) and private (railroads). Over the middle decades of the nineteenth century, American colleges found an audience for courses in trigonometry and surveying and began to offer certification programs for aspiring engineers. Rensselaer Polytechnic Institute and the United States Military Academy at West Point slowly built up pioneering programs in the broader applied sciences.[8]

With the end of the Civil War, measurable growth occurred in degree-oriented engineering education. Prior to this time American higher education, organized around a classically based curriculum, had shown some disdain for the "useful arts." The war galvanized public interest in putting the resources of the federal government towards "the application of science to the common purposes of life," and in 1862 Congress passed the Morrill Land Grant Act, providing monies for public education in agriculture and the mechanical arts. Private institutions followed this trend, and in the ensuing decade the total number of engineering schools in this country grew from six to seventy.[9]

Although most university instructors had limited time for research because of heavy teaching loads, in the last quarter of the century many academic departments embraced the idea of performing research for industrial clients on at least a token scale. The Morrill Act, and the 1889 Hatch Act, which assigned further government resources to land-grant schools, reflected and solidified the commitment of American universities to practical work.[10] Public institutions, dependent on the legislative agendas of their home states, may have followed such a course in part for reasons of economic expediency, but private universities too crafted programs of service to industry, suggesting that a social philosophy underlay the fiscal concerns of educators. As historians have noted, an older tradition of "pure" scientific research shaped the careers of many academics trained in European universities prior to 1900. With the end of the nineteenth century came a new generation of engineers and scientists trained in American universities, and with it a distinct shift in many disciplines to a practical focus. Clearly, engineering has by its nature a more ready attachment to applied study than do the life sciences and chemistry, but all can be undertaken with some commitment to "real-life" problems of agriculture or industry. By 1904, two major centers of engineering education—the University of Illinois and Iowa State College—had adapted the model of the agricultural experiment station to create engineering experiment stations. By 1927, thirty states had land-grant universities with some type of engineering experiment station in operation or under development. Other states followed suit, garnering financial support from state governments and industrial patrons.[11]

For those academics bent on building their institutions through service to industry, study of the strength of materials seemed a specialty well worth developing. As an organized system of assessing and controlling the materials of manufacturing and construction, the discipline carried the promise of immense importance to a growing economy, and thus of security and prestige for its practitioners. The foundations of such a scheme were already in place. Belief in the importance of materials analysis to industry had been gaining a following for several decades. As early as 1836 a German-trained chemist opened a consulting laboratory in Philadelphia for the analysis of iron and coal samples. The growth of manufacturing enterprises after the Civil War called for the efficient control of large-scale production processes, and raw materials and completed products became subject to scientific inspection. Gauging, sorting, and testing achieved an unprecedented importance for industrialists wishing to avoid losses that rapidly accrued in fast-paced, large-scale manufacturing operations. Metallurgy, after the introduction of the Bessemer process in the 1860s, made particularly extensive use of materials sciences, relying on sophisticated testing to control the quality of inputs and gauge the quality of outputs. This work extended from solving the problems of producing structural steel itself to controlling its application in new high-capacity engines and related machinery.[12]

Investigations might be directed towards understanding a new material or an old material being used in a new way. After 1900 the majority of tests performed for industry were "acceptance tests," which centered on the problem of whether a specimen matched performance criteria outlined in a specification or other contractual agreement. Materials testing could be undertaken by producers of a material (such as makers of steel or cement), purchasers of that material (such as construction firms), or the end-users of goods into which materials were incorporated (in these cases, architects or building buyers). Industries of each type could take one of several approaches to accomplishing the work of analysis, depending on their resources. The range of scientific tests performed on industrial materials after 1900 was large and potentially very expensive. In addition to chemical analyses and electronic or x-ray study of specimens, many destructive tests found favor. Crushing, pulling, twisting, and bending all offered evidence of a material's or part's durability, and the exertion of such forces—often on many specimens over the course of a project or production process—could clearly be quite costly. Thus, only very large businesses established in-house laboratories for truly comprehensive testing, following the precedent set by the Pennsylvania Railroad, which hired the noted chemist Charles B. Dudley to establish such a facility in 1875. This kind of laboratory might be dedicated to solving only quality-control problems or to addressing product improvement, innovation, and fundamental research issues as well.

Smaller businesses and field-based operations such as construction and mining conducted a narrower range of acceptance tests in rudimentary on-site facilities and commonly turned to university laboratories for more extensive investigations. In many locales the university laboratory provided a testing venue that was accessible, flexible in its capacities, and usually quite advanced in both equipment and technique. Such schools also served the growing world of trade organizations that sought to gather and disseminate testing data for industry after 1900. The American Society for Testing Materials (ASTM) and more narrowly focused groups such as the Society for Automotive Engineers and the American Concrete Institute exemplify this trend. Some of the organizations performed research, while some, including the immense and diversified ASTM, collated data produced externally. And as the efficacy of materials testing for industry became more widely celebrated, the demand for college-trained engineers equipped to perform tests in factory or field also grew, in turn inspiring development of materials curricula. In many ways productive businesses in the first twenty years of the century found that university engineering departments offered ideal sources of testing expertise.[13]

The Creation of University Laboratories

American universities first began equipping laboratories for the testing of materials in the 1870s, when Columbia University and Tufts College built facilities for hands-on instruction in materials science and applied mechanics. The study of materials had emerged in the 1820s as a university subject for American engineering students at the United States Military Academy. Rensselaer Polytechnic Institute introduced strength-of-materials courses in 1849, but in this period the teaching of this discipline remained limited in scope and qualitative in its approach. Beyond a handful of researchers working primarily on the behavior of metal-truss bridges, the focus of materials science at this time was on the behavior of wooden beams and columns, which, according to a history written by University of Illinois engineering professor Jasper O. Draffin, were still "proportioned mainly by custom and experience."[14] The expanded use of Bessemer steel after the Civil War, however, coincided with the impetus to technical education provided by the Morrill Act. The field grew rapidly to include the study of iron, steel, and eventually most materials of importance to American manufacturers.[15] Thus, by the late 1880s and early 1890s a number of universities, including the University of Illinois, the Massachusetts Institute of Technology, and Rensselaer, had committed their resources to substantially larger laboratories. Between 1890 and 1910, smaller schools followed, developing curricula for teaching materials sciences and building their own laboratories.

Cooperative programs, initiated by that of the University of Cincinnati in 1907, could further institutionalize this service role for academic departments. Shortly after the school's co-op program began, the University of Cincinnati's materials laboratories were made an official municipal bureau of materials testing and research. Students in the six-year program brought specimens from the commercial or municipal shops in which they worked to test in the classroom, or they assisted professors in conducting tests for other local industries.[16] Among American universities, instruction in the behavior of materials was variously categorized as an aspect of civil engineering, mechanical engineering, architecture, or subspecialties of these fields such as municipal, sanitary, or railroad engineering. Some universities included materials courses in all of these curricula. Rarely was instruction in materials testing confined simply to students of theoretical and applied mechanics, although instructors might have been based in that field.[17]

The physical facilities for testing created by many universities were impressive, and they were based on the strategic enlistment of industry both for short-term endowment and as long-term testing clientele. The engineering faculty of Lehigh University, for example, clearly recognized the importance of their location in Bethlehem, Pennsylvania, in the heart of the Lehigh Valley, one of the centers of Portland cement production in America early in the twentieth century. Their cement research facilities were of the highest caliber, and the university's efforts contributed greatly to the technical understanding of cement among local industries.[18] The habit of working closely with industry had been established in the 1880s, when Lehigh instructors turned to Lehigh Valley shops, mills, and mines for equipment with which to teach engineering specialties that had no academic precedent. Initially this arrangement was necessary for the survival of the university. Ultimately it became quite profitable, as became evident soon after Lehigh began practical instruction in materials testing in 1892. The outreach effort brought revenue from the growing steel and cement industries. In 1907 western Pennsylvania iron and steel magnate John Fritz donated to the university a testing laboratory of some 14,000 square feet, divided among facilities for studying hydraulics and the behavior of materials in general and that of cement and concrete in particular. The laboratory was equipped with state-of-the-art testing machinery at a cost of $36,000. To give some sense of this expenditure, we might note that of twenty-six university testing laboratories operating at the same time, only four others held equipment exceeding $30,000 in value. Most had been equipped for less than $12,000. Lehigh's laboratory held a universal testing machine, able to perform tension, compression, and bending tests, that could apply 800,000 pounds of pressure and test reinforced-concrete specimens up to 25 feet long. In both respects it exceeded the capacities of many commercial laboratories. The university's expenditures on cement research were rewarded by a consistent stream of commercial interest and funding.[19]

From the last quarter of the nineteenth century onward, the University of Illinois had one of the country's most ambitious programs in materials research. As early as the 1870s one physics instructor had his students build apparatus to conduct experiments in the flexure of wooden beams and the elongation of wires. In 1885 Arthur Newell Talbot, who was to become a major figure in concrete testing, began organized instruction in materials sciences in the university's College of Engineering, using homemade machinery.[20] Talbot saw in the field of concrete testing an avenue for the advancement of his own department; he installed the first, small testing machine in the college in 1888 for the study of industrial specimens and arranged to have his students' work displayed at the Chicago Columbian Exposition in 1893. He pursued simultaneously an enhanced role for materials testing in the college curricula, an expanded facility for testing, and a growing service role to the concrete trades.

Throughout the 1890s engineering students constituted from one-third to one-half of the university's entire enrollment. The College of Engineering built up its Department of Theoretical and Applied Mechanics as a sort of service department for civil, mechanical, electrical, and hydraulic engineering curricula. The department's testing equipment included several machines with loading capacities of 100,000 pounds and a large collection of extensometers, scales, and other measuring devices. Because of the success of the department with these facilities, the Illinois legislature appropriated funds for a new Laboratory of Applied Mechanics, occupied by the department in 1902. Professor Talbot obtained for this laboratory additional state-of-the-art equipment, including, in 1905, a vertically oriented testing machine that could exert 600,000 pounds of force. This was the largest vertical testing machine in existence.[21]

Between 1900 and 1930 the College of Engineering at the University of Illinois thrived as a center for the development and use of new testing machinery. A continuous program of building allowed the introduction of testing machines of unprecedented size, a number of which were designed by the college's faculty. After 1903 the development of materials testing at Illinois occurred largely under the auspices of the engineering experiment station, created by the university trustees as a research organization within the College of Engineering. (The station was the primary site in which the school's faculty consulted to industry and will be discussed below.) In 1929 the university erected a massive new materials-testing laboratory, eventually named in honor of Talbot, with an entire wing devoted to the fabrication and testing of full-scale concrete specimens. Equipped with innovative temperature-controlled "moisture rooms," the new building made possible the precise simulation of actual building conditions.[22]

In many ways we may wish to understand the commitment of Illinois to testing as a product of Talbot's own focus on industry. Having received his undergraduate education in engineering at Champaign-Urbana in the late 1870s, Talbot cultivated a lifelong interest in the expansion of the school (he was hired as an assistant professor in 1885, moved up to become head of the Department of Theoretical and Applied Mechanics, and retired in 1926). But his particular understanding of an industrial mandate for his department reflects also his participation in professional associations such as the ASTM and the American Society of Civil Engineers (ASCE). From 1904 to 1916 he served on the "Joint Committee on Concrete and Reinforced Concrete" sponsored by the two groups, and he was president of the ASTM for two years. In those roles he would have worked closely with representatives of business.[23]

A single extremely energetic and long-lived figure with such a commitment to industry seems to have been a prerequisite for the establishment of sizable testing programs in universities. With virtually identical timing, Iowa State College (ISC) also developed an engineering experiment station and an ambitious program for the testing and refinement of industrial materials. Like that of the University of Illinois, ISC's engineering experiment station came into existence through the efforts of a few highly entrepreneurial individuals, most notably Anson Marston and his staff. Marston, trained as a civil engineer, worked at the university from 1892 until his death in 1949 at the age of eighty-five, serving as dean of engineering for twenty-eight of those years. During his tenure he created a broad array of new programs in civil and mechanical engineering, often deliberately following the successful model of the Illinois faculty.[24] Marston first pursued support for an engineering experiment station at ISC by cultivating "the clay interests of the state," which he correctly believed would "cordially and effectively support an appropriation such as asked."[25] The stations' first bulletins, issued in 1904 and 1905, reported the work of Marston and others on the design of sewage systems, but data on tests of brick and other paving materials soon followed. Cement attracted Marston's attention for its double utility in road building and in the construction of silos, tanks, and other farm structures. Clearly, he understood that the admixture of industrial and agricultural service would attract the support of Iowa's legislature. A new Engineering Hall, completed in 1903 at a cost of $220,000, held separate laboratories for testing masonry and cement alongside metallurgical and electrical power generation laboratories. The Cement Laboratory held facilities for conducting simple tests in large numbers and also featured costly 50,000- and 100,000-pound testing machines for more specialized researches.

Marston maintained his interest in high-tech equipment throughout his career, often obtaining funds for new machines from university and state sources.[26] He conveyed his enthusiasm to Herbert J. Gilkey, a junior colleague and assiduous biographer of Marston, who continued to build Iowa's testing programs through the 1930s, 1940s, and 1950s. Holding degrees from Oregon State, MIT, and Harvard, Gilkey began his teaching career under Arthur Talbot at Illinois in 1921, but upon marrying Talbot's daughter in 1923 Gilkey was forced by antinepotism policies at the university to find other employment. He moved to Iowa to found that college's own freestanding Department of Theoretical and Applied Mechanics. Like Talbot and Marston, Gilkey cultivated an active role in such nonacademic organizations as the American Concrete Institute and the ASTM. His outreach to Iowa industrialists was as sophisticated and successful as Marston's had been, and his abilities to persuade his employers of the department's utility helped him build dedicated cement "moist curing rooms" and other state-of-the-art facilities at the university (figs. 1.1, 1.2).[27]

As public institutions, the University of Illinois and Iowa State College profited by proving themselves to be indispensable to their state economies. Marston was particularly adept at crafting conspicuous programs—such as the publication of *Iowa Engineer,* a popular student-run magazine that promoted private and public engineering projects—that were mutually beneficial to local industry and his own department. The wide dissemination of the engineering experiment station bulletins, while extremely useful technically, had this promotional function as well.

The entrepreneurial spirit of Talbot, Marston, Gilkey, and others in state universities was matched by materials experts in private universities. The University of Pennsylvania, for instance, created facilities of similar dimensions to accommodate the booming manufacturing and transportation industries of Pennsylvania and the Northeast. The engineering departments of the university were established in 1872 as a section of the scientific department and included geology and mining as well as civil and mechanical engineering. Two years later alumnus John Henry Towne endowed the Department of Science, which became the Towne Scientific School. In the 1890s electrical engineering was designated as a distinct area of study at the school, and the electrical and mechanical engineering departments moved to a new building. By 1906, when a fire destroyed this building, there were six hundred engineering students and a lavish new facility already prepared to house the expanding School of Engineering. Built at a cost of $1 million, the new laboratory building was the largest on campus, and its scale expressed the confidence of the engineering faculty in the importance of its work.[28]

The new Towne building provided 128,000 square feet of work space, all heated and illuminated by the latest methods. In addition to steam, gas, water, and electrical laboratories for civil and mechanical engineering courses, as well as a museum and elegant lounges for students and faculty, the building held the Lesley Cement Laboratory. This 1,700-square-foot facility was equipped through a donation from Robert W. Lesley, a one-time student at Penn's College of Engineering and a prominent Philadelphia cement merchant.[29] Lesley was also a publisher of literature on cement and an advocate of scientific research that could benefit the cement industry. His donation assured that Penn's laboratory contained the newest testing

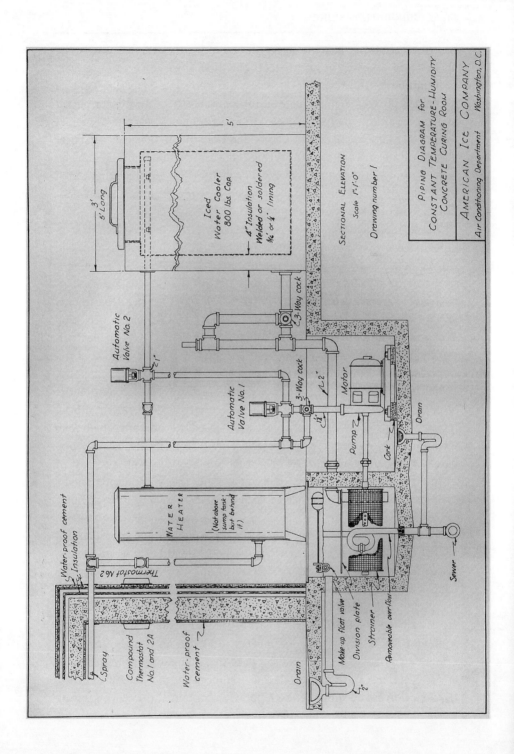

machines for tensile, compressive, and bending tests; apparatus for making concrete specimens; immersion tanks; and damp closets. Using this equipment and the testing machines located in the adjacent general materials-testing laboratories, Penn's engineering faculty took on an ambitious program of instruction and testing for a commercial clientele. They addressed the needs of area cement producers and the many concrete building projects associated with manufacturing and railroad expansion from Baltimore to New York City.[30]

The instructors most responsible for the planning and early development of the Lesley Cement Laboratory were Edwin Marburg, head of the engineering department, and H. C. Berry, professor of materials and construction. Marburg oversaw every detail of the new facility, from soliciting competitive bids on laboratory furniture to supervising the design of a 400,000-pound testing machine. He worked closely with the Riehle Company, a major manufacturer of testing machinery headquartered in Philadelphia. Berry also took advantage of Riehle's proximity—and that of the rival Tinius Olsen Company, also of Philadelphia—to design testing devices from scratch, some of which were subsequently marketed by the manufacturers under Berry's name. Between 1900 and 1915, Marburg and Berry created at Penn one of the preeminent testing facilities on the East Coast.[31]

The Dual Character of Materials Instruction

The environment that engineering instructors established for the study of industrial materials produced a cohort of young men superbly qualified to find industrial employment. When American industrialists and civil agencies implemented systematic testing and inspection procedures after 1900, they hired for the purpose young men who had been trained by university faculties in engineering and materials science but who were still junior enough to accept salaries lower than the fees their senior colleagues or professors commanded as industrial consultants. Students commonly took

---

FIGURE 1.1. *(opposite)* Schematic drawing of a concrete-curing room proposed for the Iowa State College Department of Theoretical and Applied Mechanics by the American Ice Company, 1932, by order of Professor Herbert J. Gilkey. The college's concrete-testing facilities had been among the nation's most advanced since their establishment by Professor Anson Marston in the late 1890s. *Iowa State University Archives, Gilkey Papers*

FIGURE 1.2. Cement-testing machine made by the Southwark-Emery Company in the early 1930s, shown here equipped with an automatic stress-strain recorder that rendered test results in graph form. Such "autographic" recorders were said by the company to be "virtually foolproof," and they attained increasing popularity among university materials programs through the 1920s and 1930s as less regimented instructional methods gradually lost favor. *Iowa State University Archives, Gilkey Papers*

summer jobs in inspection while still in school and moved upon graduation into long-term positions. Some worked directly for manufacturers, building firms, or city sewer and street departments, while others worked for engineering firms that transferred them from client to client. A small segment of this young group of students and recent graduates stayed in the arena of testing and inspection, eventually to become senior or supervisory personnel, but most moved after a few years into design or engineering management positions. Entry-level engineers or slightly more advanced practitioners held the majority of testing and inspection jobs.[32]

The fact that many industrial and government inspection jobs were filled by young men with short employment histories suggests that this work did indeed lend itself nicely to routinization; it did not require the presence of the most accomplished experts in mechanics or materials. But *some* experience was almost always required of those hired for testing or inspection positions, and it appears that employers preferred that candidates' exposure to materials testing and inspection be academic rather than purely commercial in origin. It was college-trained men who got these jobs, rather than self-taught applicants or wholly inexperienced men. Employers deemed classroom and laboratory exposure necessary for quality-control personnel, despite the trend in quality control towards instrumentalization and other automating techniques and the general movement in industry away from advanced training for mass-production personnel. What, then, was the content of the engineering curricula that rendered engineering graduates the primary candidates for industrial employment of this kind?

In some respects the American educators who built extensive facilities to test industrial products, and thus offer students hands-on experience with "real-life" problems, were following a German model. In the 1850s the Academy of Mines at Freiburg instituted practical courses in mechanics and machine design. Students verified principles of statics, dynamics, and strength of materials on wooden models in what are believed to have been the first cases of students working experimentally in the discipline. Other German polytechnics followed suit, presenting a trend that contrasted with the French emphasis on mathematical theory and that moved to the United States when American scholars who had been educated in Germany returned home.[33] Worcester Polytechnic Institute even erected a small manufacturing plant of its own in 1868. An interest in participatory laboratory work for students that emphasized experimentation and the solution of design problems was emerging at the same time. In 1861 the Massachusetts Institute of Technology created its school of "industrial science" expressly to train students for "practical professions." A significant aspect of the MIT program was its use of individualized laboratory instruction in the physical sciences, during which students were required to perform experiments and issue laboratory reports. The Stevens Institute of Technology opened in

1871 and three years later established the first engineering laboratory curriculum specifically designed to address the needs of industry.[34]

All of these precedents inspired faculty at Illinois, Iowa, and Penn to pursue enhanced testing facilities for cement and to build coursework around their state-of-the-art equipment. But many engineering departments bent on supplying testing services to the commercial sector created a distinct dual character for the discipline to prepare students for the broader conditions of commercial building in America. A trend towards greater control and routinization of students' classroom experiences interwove with a clear effort to claim for college-trained engineers a broad, almost intuitive—and thereby unique—expertise in the use of materials. This blending was deliberate and was probably responsible for the solid occupational identity achieved by testing personnel in commercial construction. As we trace first the rote features of engineering education and then the subjective features, we will see a remarkable compatibility and social utility to the combination.

First, we can document the trend towards greater control of student work—the introduction of constraints on students' testing procedures and problem choices. In the broadest sense this face of engineering education is reflected in the existence of set curricula: series of required classes for particular majors and degrees, and within those curricula prescribed series of activities. In all engineering departments a regimen of exercises kept the student moving through a predetermined set of techniques and in the realm of measurable achievement. In 1912, engineering students at Lehigh had to perform fifteen exercises on mechanics of materials in their junior year, including tests of steel, iron, and wood, and eight exercises on cement and concrete in their senior year. These exercises covered the range of procedures called for in any commercial application of concrete: tests on the fineness and specific gravity of cement, molding and testing cement briquettes and cubes, and molding and testing full-scale reinforced-concrete beams and columns for compressive and tensile strength. At the University of Illinois at this time, nineteen such tests were required in a single term of applied mechanics.[35]

We can also look below this level to see the degree of determinacy present in the actual content of student exercises. Refinements to existing methods of practical instruction after 1900 included the development of extremely detailed laboratory instructions, which could be so precise as to dictate the amount of time a given measurement or computation was to take.[36] Preprinted report forms, on which students would record the procedures and results of experiments, also came into common use. The advantages claimed for such forms included completeness, a saving of time, and the fact that they provided to the students when finished "a reference book which is correct, neat and of value after graduation." Professors in Penn's engineering department printed huge runs of standardized laboratory report forms and distributed them both at Penn and at several other universities that expressed an interest.[37]

The report forms could help instructors exert extremely close control over students' conduct in the laboratory. At Iowa, instructors in materials testing used preprinted forms that closely resembled those used in commercial cement-testing operations, and it is clear that the process of filling out the forms was itself considered a matter of training and discipline. Students were divided into squads of five, with one student assuming the role of "recorder"; the others acted as "operator," "observer," "computer," and "helper." At the top these forms called for exact information about when and by whom an exercise had been undertaken. The unlabeled columns below were filled in by students according to the nature of the individual test, but an accompanying syllabus firmly instructed the recording student that "before any data are taken he should secure instructor approval for column headings." What's more, the recorder was admonished that "under no circumstances are observed data to be recorded elsewhere and transferred to the report form." Iowa's instructors conceived of the reportorial aspect of testing as one requiring close supervision and habits of accuracy (and honesty, perhaps), both of which evidently could be derived from such constraints.[38]

The development of automatic and autographic (self-recording) testing machines was also part of this trend. In using such equipment, students would increase loads on specimens incrementally by mechanical means. The machine automatically recorded the breaking point of a specimen, thus freeing students from such interpretive tasks as setting or reading scales. By 1925 some machines automatically rendered readings into stress-strain diagrams, indicating in one image the degree of stretch, compression, deflection, and twist detected in a specimen.[39] Such innovations speeded testing work, reduced the probability of error, and may have protected the substantial investments that such machines represented. Watching students struggle with expensive machines prompted some instructors to build failsafe mechanisms into laboratory equipment. Edwin Marburg, working with the Riehle Company on a customized six-speed machine for Penn's laboratory in 1905, designed a special "interlocking mechanism" that would "make it impossible to throw the machine into more than one speed at a time." This, he explained, would prevent breaking of the speed gears.[40] Smaller instruments were also susceptible to refinements that reduced the need for student discretion while speeding up the testing work. Irving Cowdrey and Ralph Adams, two professors of materials testing at MIT, praised in their 1925 textbook the development of electronic micrometers attached to buzzers. With such tools the student received instantaneous and "exact" indication of when a preset measurement had been attained.[41]

The pursuit of control over students' work did not always require the introduction of instruments of increased technical complexity. For example, Cowdrey and Adams applauded shortcuts in testing such as permanently setting one arm of a caliper. This incapacitated part of the device but relieved the student of having to check "the coincidence of both parts of the divider" with each measurement. New, quite simple apparatus could also bring about this limitation of student discretion. The "flow table," introduced around 1905 for the determination of wetness in a sample of cement, is a case in point. At this time, Vicat and Gillmore needles, both of which reduced the determination of cement's wetness to a calibrated scale, offered the most established tools for this purpose. These appear on many laboratory inventories and in virtually all concrete textbooks of the day. They are relatively simple and inexpensive devices based on a controlled dropping of a needle into a small cup of wet cement. But the needles have the limitation of registering progressions in hardness only once a certain degree of hardness has been reached (if a specimen is too fluid, the needle simply sinks completely). The flow table, an even less complex device, allowed the fluidity of even the wettest specimens to be calibrated. An operator dropped a quantity of wet cement or concrete from a given height onto a surface marked with concentric rings. The fluidity of the specimen was indicated by the size of the circle it covered. Students now had a more "sensitive" instrument, able to measure a wider range of conditions, that expressed results in quantitative terms.[42]

A sense that students' work had to bring measurable results pervaded discussions among engineering instructors in this period. With the notion of "doability" paramount, many exercises would be put in the context of problems borrowed from the commercial world. Juniors in Penn's engineering program of 1909 were required to write "summer memoirs" on an engineering work or manufacturing plant. One popular course on power plant design and operation, published in textbook form by Penn instructors of mechanical engineering, required students to measure the costs of operating and maintaining plants. Typical senior thesis titles in Penn's engineering school of the same period include "Design of a Reinforced Concrete Building for Light Manufacturing" and "Design of a Reinforced Concrete Building Adapted for a Cotton Mill." At Iowa, prizewinning undergraduate papers, judged by officers of the Iowa Cement Users Association for a 1908 competition, included "Tests of the Impermeability of Reinforced-Concrete Pipe" and "The Improvement of Cement Mortar by Grading the Size of Sand." Both papers were based on original research at ISC's engineering experiment station. Similarly, by 1914 MIT had divided engineering subjects into such specialties as factory construction, foundations, and refrigeration, indicating the school's emphasis on results-oriented instruction.[43]

Such assignments conveyed to students the nature and pressures of commercial work, and much of this pedagogy centered on a perception that undergraduates were immature and required close management by faculty. W. K. Hatt, a prominent materials instructor at Purdue, made it clear that strict limits on undergraduate discretion were vital to successful thesis completion. While it might be appropriate to let a student choose his thesis topic and "details of apparatus," "it is thought that it is better that he should be given sufficient guidance and should thus come to the end of the year with some definite conclusion reached, than that his energy should result in a mass of tangled data such as the average undergraduate obtains by his own inexperienced planning."[44] And yet alongside this promotion of closely watched and highly regimented student work, there existed in the minds of many engineering instructors a clear sense that overdirecting student efforts posed dangers to the effectiveness of engineering education, and to no less a treasured objective than the status of engineering professions. Wickenden declared that individualized laboratory training was one of the significant innovations of American engineering education, but for all the emphasis they gave it, instructors believed that serviceability must not displace students' development of discretionary and observational skills. In fact, serviceability was in part constituted of just such attributes, routinized and instrumentalized procedures employed alongside barely defined techniques based on the subjective talents of the operator. With such a doubled design for their discipline, materials instructors hoped to create a compelling identity in the world of commercial construction.

Support for both types of work often issued from the same faculty members, with the most common recipe for success being a blend of direction and discretion. Categories of work such as "practical" and "problem-oriented" were as easily mustered in the service of developing students' discretion as they were invoked to add direction. When a former student wrote to University of Pennsylvania professor H. C. Berry with a question about the behavior of concrete railroad beds, Berry gave only a cursory reply. He directed the writer to "investigate the problem experimentally," at the same time advising, "I am glad to note that you are up against practical problems." Here practicality implied an occasion for independent thought.[45] Anson Marston, who had brought a massive, state-of-the-art testing laboratory to his department at ISC, declared in 1904 that thesis topics there were always "of importance to industry [but] this object is not furthered to the detriment of the student's interest to whom the thesis must be first of all an opportunity to think for himself and to apply principles previously inculcated in the regular course of his studies."[46] In a textbook intended for engineers seeking careers as supervisors of construction projects, Daniel Mead of the University of Wisconsin wrote in 1916, "The essential aim of technical education is not so much to impart technical knowledge to the student as to furnish the training which will enable him to understand and investigate the conditions which surround a problem . . . to ascertain and analyze the elements which influence or modify it."[47] The ideal engineering mind seemed to be a complex mix of scientific method, training in the use of instruments, and raw ingenuity.

This project of inculcating, or perhaps identifying, ingenuity in students of materials shaped professors' approaches towards the material bases of instruction. Many engineering faculty believed that laboratory instruction must not, as one civil engineering professor put it, "degenerate into formal compliance with instructions [or] be taught with such definite direction that the work becomes mechanical."[48] Some instructors of materials testing felt that it would be unwise to provide students with preprinted forms for their laboratory reports. One professor at the University of Illinois who disdained forms claimed that although his approach made more work for the professors who had to grade laboratory work, and slowed down courses because students made more errors than they would otherwise, "any loss in the amount of ground covered is more than offset by the training the student obtains in developing ideas, assuming responsibilities, and gaining confidence in himself."[49]

The growing presence of machines and the trend towards automation

FIGURE 1.3. Students in concrete construction class, Iowa State College, ca. 1909. Most university engineering programs of the period combined hands-on work with theoretical instruction, claiming that the dual nature of such training uniquely qualified university graduates for industrial employment. *Iowa State University Archives, Marston Papers*

worried other instructors. One recommended that students in materials classes start by conducting tests on hand-powered testing machines rather than on motor-powered ones. In this way, students would develop "the correct mental appreciation" of the physical forces involved in materials sciences. Cowdrey and Adams of MIT maintained that many types of testing apparatus were, for teaching purposes, unnecessarily complicated. W. K. Hatt at Purdue, who encouraged close control of some features of student work, believed that it would be "of no particular advantage to have automatic or autographic testing machines for student use." Instructors teaching the materials-testing laboratory course at Iowa in 1935, who assuredly knew the value to students and industry of cutting-edge technology, warned that in certain instances hand work offered the chance of greater precision in measurement than did reliance on mechanical devices (fig. 1.3).[50]

Machines were elsewhere defined as means of encouraging independent thought—perhaps not surprising, since they were achieving increased popularity in university testing laboratories. At least one professor fretted that too much preparation was being done for students in the interest of efficient use of laboratory time, so that the teaching of rudimentary knowledge, such as how to turn the machines on and off, was being neglected. And Anson Marston, ever the strategic planner, protested in 1906 to the president of Iowa State College that as long as his engineering students were forced to share testing equipment, he could not provide an ideal training ground for them. Apparently even close-at-hand witnessing of another's work did not offer a sufficiently instructive laboratory experience.[51] These seemingly contradictory ideas about machine-based instruction were actually not incompatible. In general, the benefits of laboratory work for testing students included cultivating powers of observation and expression, mental efforts that one professor contraposed to the presumably unreflective process of "simply seeing the specimen break." Even shock and consternation, clearly emotional rather than intellectual aspects of technical practice, seemed to have value. A professor of civil engineering at the University of Nebraska claimed that difficulties in machine operation brought students to a state of "surprise . . . one of the first psychological requisites of good teaching."[52]

As will become clear in later chapters, university instructors in the years between 1900 and 1930 were echoing in many ways the sentiments of the industrial employers to whom their students would ultimately turn for work. We can note for the moment that Charles Dudley, the industrial chemist, told an audience of engineering instructors in 1906 that as far as training in materials testing was concerned, the "behavior in service" of materials was the best type of test (as opposed to the necessarily artificial conditions of testing by machine). But while praising authentic experience for engineering students, he warned against excessive practicality, overemphasis on "methods, manipulations . . . and accumulated information" to the neglect of "underlying principle and reason." To be "real," the student's experience needed both a concrete (so to speak) and an ideal component.[53] Materials instructors defended the essential artificiality of laboratory work on this basis.

The Role of Theory in a Formulating Discipline

As the close examination of standards and specifications for concrete in chapter 2 will show, the abstraction of knowledge about materials became fundamental to the discipline's identity. But on the level of course design itself, Dudley had touched on an important point. In their efforts to promote intuition and "original thought" over the course of practical instruction, engineering professors evinced great concern over the question of when, and to what degree, they should teach their students formulas and methods of theoretical analysis. We need to be cautious in understanding what these instructors meant by "theory." Along with "underlying principles," the so-called theoretical work often appeared under the rubric "fundamentals," which for these instructors encompassed a range of student pursuits from chemistry courses to familiarity with mathematical formulas. This category of subject matter was invoked largely in opposition to "specialization." The whole problem of "overspecialization" seemed to hinge on instructors' anxiety that engineering students were becoming bogged down in narrow technical proficiencies to the detriment of foundational knowledge. The meaning of all these terms is best understood in the context of their social utility. The instructors' nomenclature functioned at least in part to define the discipline's notion of its relationship to other scientific fields.[54]

In 1918 physicist Charles Riborg Mann produced a report for the Society for the Promotion of Engineering Education (SPEE) and the Carnegie Foundation on the state of engineering education in America.[55] He found that most schools segregated the teaching of theory or the fundamentals of science into the first two years of instruction and practical work into the last two years, a pattern that he claimed had been in place since the mid-nineteenth century: "In this matter the 1849 curriculum at Rensselaer imported a French style that has been followed implicitly ever since. The conception underlying this and all later curricula is that engineering is applied science; and therefore, to teach engineering, it is necessary first to teach science and then to apply it."[56] Mann's report called for a change to this approach, but in any case his generalization is not adequate for understanding the attitudes of materials instructors, who did not necessarily see materials analysis as applied theory.

The laboratory instruction manual for a 1904 course at the University of Illinois Laboratory of Applied Mechanics, for example, lists four objectives of laboratory work: "(a) to familiarize the student with methods of testing, (b) to give him practice in drawing conclusions from physical tests, (c) to illustrate methods of failure of materials and to show conditions under which different materials may be considered reliable, and (d) to compare theory and observed phenomena."⁵⁷ These objectives are not necessarily listed in the order of their priority to instructors, but it is nonetheless evident that the connection of theory and observed phenomena is a *compartmentalized* task. It is not the ultimate aim of laboratory work but only an aspect of it, one goal among several.

In a similar vein, an instructor of mechanical engineering at the University of Cincinnati wrote in 1912 that "most of the theory essential for the conduct and reporting of experiments is simple enough. The important thing is drill." In the publications of SPEE between 1900 and 1930, for every instructor offering advice on how to teach theory, another would voice concern that diagrams and equations did not "represent facts": "The more we bring the student into contact with the real mechanisms of physical deformation in actual materials before proceeding to conventions or abstractions, the more likely we are to reach one of the ends of education, namely, to develop the power of original thought."⁵⁸ One University of Illinois instructor in reinforced concrete even went so far as to warn that because formulas tended to obscure particular physical relations, they encouraged the student to proceed in an intellectual direction counter to the natural, which was "by progress from the particular to the general." He noted that "any other order is as confusing to the mind of the student as would be an attempt to interpret the plot of a movie drama, when the film is made to run backward."⁵⁹

The use of theory was not, of course, threatened with banishment from the university engineering classroom. It was frequently secondary in the concerns of early-twentieth-century materials science instructors but very rarely dismissed outright.⁶⁰ Some academics in the applied sciences recognized that theoretical approaches offered not only potentially inexpensive means for problem solving but also a prestigious association with "higher" sciences.⁶¹ But theory was not treated as an end in itself, either in pedagogy or as a component of laboratory work. The observation and interpretation of actual materials were dominant and carried in these conceptual schemes their own intellectual cachet.

In some ways materials instructors were laboring to devalue conventionally prestigious techniques of science and lift their own very practical activities to a realm of social significance. In one such effort, Charles Ellis, a professor of civil engineering, declared memorization of formulas to be without merit as a teaching tool. Ellis saw himself as embodying Mann's ideas on encouraging student initiative, elevating practical laboratory work above rote recitation in order to do so. He offered in 1918 the "tragic experience" of the "student Jones": "He registers in a certain course and spends three hours weekly in the lecture-room taking copious notes, listening, sleeping or perhaps reciting *words* which he has learned from his notes or text-book.... At the end of the semester, by pumping all night, he sufficiently increases the pressure on his watered stock of knowledge to be able to flush a dozen pages during a formal and final examination which is conducted with great dignity."[62] Ellis's derision is obvious. Formulas as conventionally taught were major offenders in this scheme:

> Brown ... who represents the average conscientious student ... knowing the possibility of being called upon to derive the formula upon the blackboard ... follows the algebraic transformations very carefully to the end, memorizes the formula and says to himself, "I have got my lesson."
> .... But what does he *see*—the beam? I do not think so. The formula has become his hero—and it finally becomes his master.[63]

Ellis advocated, instead, teaching the behavior of structures through graphical analysis—a means of describing the mechanics of materials with geometric notation (rather than with algebraic terms) that had existed since the mid-nineteenth century.[64] By such a method, Ellis concluded, the student would develop not only initiative, scientific attitude, and resourcefulness but also the "mental power to cope with life's problems, the answers to which are not to be found between the covers of any text-book."[65] In this last line lies the essence of Ellis's argument: effective engineering is not like scientific work in every respect, nor should it be. Where there is work (i.e., physical, technical, *productive* work) to be done, one needs not classical bodies of knowledge but "the mental powers to cope with life's problems"— to think on one's feet. Engineering in these terms comes across as an intellectual undertaking as challenging and vital as scientific inquiry, if not more so. After all, formulas can be passed along and even memorized. "Mental powers" are a far less transferable commodity.

This was occupational boundary work of some complexity. The rise of testing and inspection jobs in industry promised engineers a source of income and security, but the very efficiency they claimed—fast, instrumentalized testing procedures; a body of knowledge codified in textbooks and college courses—could undermine their claim to untransferable skills. The greater the formalization or quantification surrounding a field of engineering expertise, the greater the insistence on extratechnical proficiencies. Mann, like many others, applauded the growing interest among engineering departments in courses on cost valuation, contracts, and other business aspects of technical work. But he warned that where courses became too materialistic, the student risked sacrificing no less than his "powers of abstract thought and humanistic ideals."[66] The distinction between "information" and "education" was maintained by many educators of the day, and the popularity of this logic among engineering professors is telling.[67]

As lofty and indeterminate as such "powers" seem to be, ascribing such unmeasurable characteristics to engineering students had a significant practical function in the competitive marketplace. A professor of civil engineering and member of SPEE wrote in 1917 that "mental initiative, resourcefulness, and self-reliance are the qualities which distinguish the professional engineer from the technician or mechanic."[68] Because the work of materials testing could in large part be reduced to rationalized units, as contemporary instructional methods show, other characteristics were needed to distinguish the "professional engineer" from the nonprofessional in the marketplace. The very adaptability to routinized practice that made materials testing so suitable for commercial application could have threatened materials experts' authority, had they not so emphasized these other less tangible criteria for professional status. By attaching to the university-trained tester a particular acumen—or intuition, as the engineering instructors were fond of calling it—the educators associated tasks of materials testing with specific persons rather than with specific machines or methods.

Strategies for Self-Definition

That engineers worried about the leveling forces of commercial employment is clear. Wickenden's 1930 assessment of university engineering programs warned that if engineering instruction became too routinized, the exceptional student would be reduced to a "harried quantity producer."[69] The standardizers appeared eager to separate themselves from the standardizing process. In seeking to define themselves as special, relative to practitioners with less or different training, the materials instructors were not unusual in combining practical skills with less concrete personal attributes. In the same period, for example, engineers employed by government agencies touted their political independence, and scientists of different kinds made much of their own objectivity.[70] The materials specialists' choice of attributes, however, was geared towards their position in a system of mass production. They wanted to remain vital to the routine construction of concrete buildings and other manufactured goods, and their professional claims should be seen in light of these material conditions rather than as identical to those of other scientific disciplines.[71]

Understanding the limited role of theory in the daily work of the materials scientists is particularly important in identifying their occupational goals. Sociologist Everett Hughes has written that the strategic balancing of universal and particular knowledge is a hallmark of professions. For Hughes, this balance takes the form of an intellectual detachment that leads professionals to perform tasks with disinterested thoroughness, on the one hand, and a "deep interest in cases of all kinds"—an attraction to the universal—that leads them to "pursue and systematize knowledge," on the other. Such systematized knowledge, Hughes writes, is "theory."[72] The balance sought by the academic materials scientists grooming their students for future professional status was not between practice and theory but between standardized practice and personal judgment. Theory had little place in the day-to-day work of materials scientists, and universality translated for them into instrumentalization, instead. Daniel Mead distinguished between theory—encumbered by "limitations"—and good judgment.[73] As Cowdrey and Adams pointed out, a highly systematized approach to testing work, combined with personal discretion, could yield valuable findings: "careful comparisons" turn the results of a single test from "unattached and barren facts" into part of the tester's "general knowledge of the investigation." Their calculus for commercial success found talents of discrimination to be more useful than immersion in theoretical work.[74]

This professional recipe certainly gained for the concrete experts "a larger measure of autonomy in choosing colleagues and successors," considered by Everett Hughes to be an identifying goal of professions.[75] It is important to note also that the lack of theory in their daily work was not a limiting factor in the careers of these materials experts, either within or beyond the universities. Some historians of early-twentieth-century American engineering have seen engineers' accomplishments at this time as having been limited by the profession's focus on industrial production and design problems. These restrictions in specialty or subject matter supposedly hindered both the advancement of industry and the professional achievements of the engineers.[76] But the materials experts' refinement and rationalization of testing techniques brought greater efficiency, lower costs, and increased reliability to concrete construction.[77] As for their professional achievements, the instructors sought for their graduates a position of authority not "above" the realm of routinized industrial production and design—as research and development might have offered—but within that realm. As one professor at Rensselaer wrote of materials curricula in 1912, "The main object should be to produce men, who, because of their training, are better able to write specifications on tests or to judge the value of any test according to the requirements of a given specification."[78] This seemingly unassuming program actually called for regular employment in the ever-widening sphere of industrial production. With their construct of an experience-based yet efficient application of science to industry, the academic materials experts positioned their students quite shrewdly.

Character and Culture in the Testing Laboratory

For many engineering instructors, it was a short step from praising the frankly subjective and unrationalized techniques of modern testing to claims of exemplary personal character for the budding materials experts under their tutelage. In textbooks and educational journals, instructors supported the idea that the well-trained and efficient engineer would occupy a heightened moral position in the modern world. This concept was linked in some literature to high cultural status for their graduates—an ability to preserve and forward the finest impulses of human civilization. In part this stewardship of culture was to be enacted through the inclusion of arts and letters in the engineering curriculum, the benefits of which were seemingly unlimited. As consulting engineer J. A. L. Waddell told members of SPEE in 1915,

> The broadening of engineering courses so as to include many studies of a non-technical character is a great desideratum; and its importance cannot well be exaggerated. . . . English, a foreign language, political economy, ethics, history of engineering, oratory, logic. . . . Such studies will tend to make of the student an educated polished gentleman. . . . A general knowledge of political economy is a requisite for every deep thinker—and who has to think more deeply than the engineer?[79]

But so too might the engineer contribute to high culture. Anson Marston believed that the mechanic arts "are even essential to modern intellectual development and artistic expression and they contribute to our spiritual requirements."[80] Lest one think that technical and humanist pursuits were actually separable, he added this note: "The ideal specialist of the future will not be a narrow enthusiast, caring for nothing outside his one line of work. Rather, he will be a man who adds to a broad well-rounded general training the reverent possession of some high, special knowledge and skill, which he devotes unselfishly to the service of humanity."[81] Much of this language seems so lofty as to be merely ornamental, but we are again seeking the social utility of such literary formalities, and those utilities turn out to be substantial. We are seeing a further step in the instructors' program to associate the efficacious use of new technical knowledge with a particular social cohort.

As university engineering faculty worked to define the nature of their own expertise and attach a particular set of capacities to their students in the public's mind, they took on the parallel task of detaching those capacities from people who had not passed through the university. We can see this project as building on and supporting the function of American higher education as a mechanism of social segregation. Intentionally or not, American universities appeared to favor the advancement of white, male, and native-born students over nonwhite, female, and immigrant groups. As historian of American education Paula Fass has pointed out, leaders in primary and secondary education in this period operated on presumptions about the variable "native talents and probable future lives" of different genders and ethnic groups, and university faculty shared this ideology.[82] In the pedagogical agendas of university engineering instructors, issues of character and cultural endowment fulfill this segregational function. Moral and cultural standing are not measurable quantities in any immediately obvious sense. Unlike some calibrated level of, say, visual acuity or manual dexterity, these attributes are not readily provable. In this regard they were ideally suited as markers of "innate" and "necessary" talents among aspiring engineers. The materials experts, outlining requirements for success in their field, attached moral and cultural achievement to a set of personal features—namely, masculine gender and other "inherited" characteristics. With this kind of attachment one could formulate an image of good engineers as best derived from a "stock" of white males. If gender and ethnic origin can be seen to determine technical prowess, one then has a means of attaching technical ability, and all its associated social rewards, to that social "stock."

We should recognize that on one level, by stressing their role as moral or cultural instructors, university engineering departments may have relieved themselves of some financial burdens. In declaring the university an important site of training in arts and letters, even for future engineers, faculties sought to leave selected portions of technical training to the students' future employers. Anson Marston, Charles Mann, directors of the Lewis Institute, and others indicated such intentions.[83] Technical instruction cost more than classes in language or literature; laboratories for state-of-the-art engineering training were particularly costly. By delegating some of that practical instruction to industry on the premise of retaining a more cultural role for themselves, engineering faculties may have justified limiting their expenditures without incurring the criticism of industry.

In another approach that gained engineers a cultural cachet with little outlay, university engineering faculty identified the subjective features of engineering training as a means of integrating character development into their teaching. No need even to institute arts and letters classes if one sees engineering itself as a means of societal uplift, as did John Price Jackson in summarizing for other educators the cultural accomplishments of modern engineering programs:

> Are not the scientific, or better still, the engineering ideals of the day the embodiment, though imperfect, of these thoughts and ideas which have been slowly developing and growing for so many decades or even centuries? . . . We all believe the educated man should have judgment; . . . should be able to perform his share of the social duties of humanity; should have . . . so true and avid a desire for knowledge that throughout life he will gather about him the writings of his great contemporaries and those who have gone before.[84]

It does not denigrate the sincerity of Jackson's altruism to note that along with the improvement promised to civilization, the growth of engineering would also benefit its practitioners. Charles Dudley, the successful industrial chemist, informed engineering educators in 1906 that "'The Testing Engineer' is destined to play continually a more and more important part in the development of civilization," and one wonders why anyone in his audience would have disagreed. In a somewhat more subtle vein, Anson Marston compared a new six-year engineering degree at Iowa State College with curricula of law and medicine, a fairly common type of comparison among instructors in technical fields.[85] In all such rhetoric, engineering comes across as a reasonable substitute for such conventional cultural subjects as arts and letters.

But if there was a certain institutional expedience to such claims—allowing university engineering faculty to give a broader utility to their curricula and thus a firmer fiscal justification—many aimed at a larger social purpose. A careful examination of the pedagogical programs outlined by these instructors reveals a number of ways in which general language about engineering in culture is accompanied by specific prescriptions for the role of engineering graduates in the modern workplace. Many engineering instructors engaged in this exchange clearly regarded the competitive conditions of employment awaiting their students with great concern. A carefully nuanced, but recognizably ambitious, program emerged to exclude non-university-trained practitioners from the upper reaches of industrial work.

Engineering for Leadership

The essence of this campaign was that positions of social leadership for engineers could be enacted in the very organization of technical education. First, we can examine the general composition of curricula as it fulfilled this separatist function. In 1915 J. A. L. Waddell assured a SPEE audience that the control of high-level industrial employment by college graduates represented a "natural and correct hierarchy." As he described the most efficient means by which universities might supply engineers to industry, a subtle but strategic attempt at gatekeeping unfolded. Waddell found it regrettable that in order to meet the demands of industry, universities must turn out "a large number of effective men each year" for the United States to "succeed in foreign trade," because in this rush to produce engineers, teaching is reduced to suit the least talented student instead of "the man of superior ability and quicker perception." Waddell proposed an enhanced system of polytechnic and trade schools for less promising students. Young men showing real ambition in these lesser schools might then, Waddell says, be offered further university-level training at no cost. This seems at first glance a plan likely to expand access to college engineering programs. But Waddell also observes that in any occupation there "must be many hewers of wood and drawers of water," and in his plan neither the polytechnics nor the trade schools "should be obliged to spend their time on cultural study." Waddell clearly associated arts and letters with high-level engineering-career concerns. While his proposal does not directly preclude the possibility of a trade-school graduate ascending to university graduate, it discourages that outcome by omitting for the trade school a hallmark of elite engineering: cultural and ethical development.[86]

That these constructs were consonant with prevailing ideas about society and work after 1900, and not merely the hopeful self-aggrandizing rhetoric of aspiring elitists, becomes clear when we look at ideas promulgated by somewhat less prestigious schools. Directors of the Lewis Institute in Chicago, founded in 1896 to offer high school and college-level technical training to middle- and lower-income students, reiterated a familiar hierarchy. Their promotional materials present the image of a stratified American workplace filled with men of exceptional abilities ("captains") who establish companies and earn at the highest rates; a larger group of men of midlevel talents ("lieutenants") who "design machines and bridges, and supervise factories" and receive midlevel salaries; and finally the still larger group of "sergeants, corporals, and privates" whose jobs remain unspecified but who will earn the lowest wages of the three groups.[87]

The association of cultural uplift with occupational leadership in technical fields is further indicated by its deliberate omission from curricula intended for less elite populations of engineering students. In a 1903 bulletin the Lewis Institute lavishly praised the liberal arts components of engineering programs at MIT, Cornell, and Columbia University. We might expect the institute to claim a similarity to those august institutions, but the authors take a tack that both justifies their own lack of liberal arts curricula and asserts that correlation of economic standing and occupational potential: "To *encourage* engineers to take a large amount of general collegiate study is in accord with the spirit of thoroughness. To *require* it, however, would be to lengthen the course beyond the slender means of many promising men. . . . To require it would therefore be an aristocratic measure utterly foreign to the spirit of the Institute."[88] A sense of condescension is inescapable.

Next we can look with greater specificity at the actual nature of engineering coursework and highlight the discretionary nature of training in materials testing in this hierarchical context. For many observers the well-trained engineering student acquired analytical skills that simply elevated the thinker above his subordinates. Frank Marvin, president of SPEE in 1901, directed the society's members to "first, last and always let students be trained to do their own thinking. And to form their own judgements; to test the statements of others by their own mental processes."[89] The habit of independent thought positioned the tester relative to other people in the workplace. When John Price Jackson said that good engineering training was "good for the moral man" because it corrected in students "undue respect for authority"—particularly in students whose "constant attitude of mind is that of submission to dogmatic authority"—he sketched the same connection of technical and social superiority that is discussed here.[90] The suggestion is not of a radicalizing self-empowerment but of able persons dispatched to take on authoritative roles in already hierarchical settings.

Often, this view of engineering attached to a linked system of technical competencies and class membership. Ernest McCollough, a professor of civil engineering, told members of the American Society of Civil Engineers in 1912,

> There is no sharp class line, as some would have us believe, between the millions of workmen who need some industrial training in continuation classes, the hundreds of thousands of specially expert workmen and foremen, who need training in a whole system of cooperative trade schools in each state, and the tens of thousands of expert engineers, superintendents, managers, business promoters, contractors and owners who need to have taken regular professional engineering courses.

And yet he goes on to reveal that in his mind there are clearly inherent differences of capacity: "The engineers, according to the great historic definition, are those who 'direct the great sources of power in nature, for the use and convenience of man.' Properly to perform this great function they must have mingled with the privates and non-commissioned officers of the great industrial army, and have acquired a true and sympathetic understanding of their needs, limitations and possibilities."[91] The disingenuous tone is not unique to McCollough. Anson Marston held in 1931 that "the engineer is the *professional brother* of the laborer, the mechanic, the technician, the foreman. He plans and directs what they do, and strives in order that they may have worthy work to do. He should always remember that he is their *brother*, but *never forget* that he is their *professional brother*, with the standards, the obligations and the ideas of a great learned profession to uphold."[92] The engineer acquires in these schemes nothing less than the moral obligation to uphold his elevated occupational status.

So far we have traced this occupational vision as it divides the world between those with engineering degrees and those without. It is important to recognize that that division often followed lines of gender and ethnicity or race. In some sense these exclusionary patterns, in which white men retained an occupational opportunity denied to other groups, reflected existing structures of professional contact and influence. For example, when applications for university engineering programs at Penn and a "Personnell Service Leaflet" prepared for Iowa's engineering graduates asked for a student's father's occupation or family religion, they pinpointed factors that defined the social world of many Americans, in small towns or large cities. Alumni records of Penn's engineering school indicate a preponderance of men claiming membership in local Episcopal churches, and this social commonality is not surprising.[93] But these social connections were not incidentally drawn along lines of gender or ethnicity. As numerous historians of scientific management have shown, the anonymity and scientific tenor of modern hiring practices did not banish sexist, racist, and antiimmigrant tendencies among employers, and in fact may have embodied those very biases. Similarly, in the hierarchical pedagogical visions of early-twentieth-century engineering faculties, gender and ethnic segregation may have been foundational. The fact that a "Scholarship and Personality Record Card" from Iowa asked in the 1930s for a student's "Color and Race" indicates that there were multiple "colors and races" represented in the student body, or at least eligible for admission. In the 1920s at least one engineering department at the college welcomed students from China and the Philippines.[94] But if integration had been established on the most profound level, such information would have been moot, which clearly it was not. (This being long before the advent of affirmative-action quotas for higher education, such information was unlikely to have been solicited to confirm a diverse student body.) It becomes apparent that this information helps support certain social structures of opportunity.

We can first briefly trace conceptions of gender in the world of engineering education. Women were rarely found in university engineering programs of 1900 to 1930, and certainly programs in materials testing were no exception.[95] Schools that offered men and women core courses in math and science added machine tooling or materials testing as "practical work" for men while assigning women to courses in "household economics."[96] When women did graduate from engineering departments, even the most ambitious saw before them narrowed occupational paths. One striking case is that of Elmina Wilson, who graduated from Iowa State University in 1892 with a degree in civil engineering. She was praised by Dean Marston for her excellent work and was hired as supervisor of the college's drafting room. But when Wilson looked forward in her career, hoping to become a surveyor or other field-based practitioner, she found little encouragement. Wilson wrote to Marston in 1904, "I have been dreaming while awake, that you are a Sanitary Engineer with headquarters in Chicago, that I am doing your drafting and perhaps a little field work. . . . Does that not make a good dream?" That her gender rather than any other aspect of her person probably rendered her ineligible for such a career path emerges in the next line: "You once said that if I was a man you might take me for some such work and I am sure there is need of a good Sanitary Engineer in the west and more money than in College work."[97]

Marston's reply does not directly address Wilson's ambitions but rather promises her employment within the university for as long as she might like. (She stayed until 1905.)[98] For Marston, the young woman's ineligibility for commercial or even fieldwork is certain. He held that there were matters of engineering beyond "knowledge of the sciences" and "experience," which he summed up as late as 1931 as the "Fundamental Manhood Qualifications of Engineers." In this gendered categorization he placed "Character" (including integrity, responsibility, resourcefulness, and initiative); "Judgement" (meaning "common sense, scientific attitude and perspective"); "Executive Ability" ("understanding of men"); and "Temperament" ("ability to make friends without yielding principles"). Daniel Mead called his own list of desirable character traits for engineers "Specifications for a Man," a feature included in the 1916 and 1933 editions of his popular textbook.[99] The president of Clark University, discussing secondary technical education, was surely not expressing a rare sentiment when he said in 1906 that women were simply "not rough enough to teach boys in high school," but the masculinizing of personal attributes was no less effective for its conventionality. To ask that laboratory work be conducted "on a high plane of gentlemanliness" is, if nothing else, to plant in students' minds an image of a single-gender workplace.[100]

Other categories invoked by engineering professionals in this period reflected additional, generalized notions of appropriate social behavior. In a 1934 memo to his fellow professors at Iowa, Herbert Gilkey considered it necessary to include among the "objectives of engineering education" features of "normality as an individual." Among these were "proper orientation with reference to religion" and "a normal family life."[101] These caveats were probably meant to identify improper conduct per se—a failure to attend church; unusual sexual orientation—rather than groups of people as such. Yet their seamless connection of technical capacity and social behavior was not accidental and in effect would serve to exclude people of unusual habits or attributes from the world of professional accomplishment. The idea of listing and calibrating personal attributes attracted many engineering professors. A. A. Potter, dean of engineering at Purdue University, developed soon after 1900 an inventory of personal characteristics, invoked by Mead's 1916 text as a tool for "acquiring the habit of observation, in the formation of correct self-knowledge, and in the development of correct judgment of others." If some of the features Potter lists, such as height and hair color, seem merely to be neutral identifiers, we might give more social weight to Potter's categories of "voice" ("Musical, well modulated, pleasant, monotonous, harsh, nasal, drawling, lisping, guttural, or falsetto?") and "accent" ("Bostonian, Yankee, Southern, Western, or Foreign?"). It is hard to imagine that these terms were not in some way normative. Even if they did not correlate with firm judgments about a potential business or social associate, they imply, like Potter's ratings of "Piety" ("Fanatical, religious, lukewarm, indifferent, irreligious, or scoffing?") and "Esthetic Taste" ("Excellent, good, medium, poor, or bad?"), that these aspects of personal conduct had meaning in business and social relations of the day. They might reflect useful information about oneself or another and become the basis for significant categorizations of people.[102]

Importantly, Anson Marston also mentions as a "Manhood" qualification "Health," which he defines as freedom from both "injurious habits" and "inherited weakness." The latter phrase is telling. It could easily encompass poor eyesight and such eugenic transgressions as venereal disease but could extend also, of course, to one's race. Like the inborn trait of gender, one's race offered the promise or prohibition of a successful engineering career. Race was routinely requested on university applications, and in the 1910s and 1920s questions about the innate educability of "nonwhite" populations proliferated with the immigration and in-migration that filled American cities with people of color.[103] C. C. Williams, dean of engineering at the University of Iowa, wrote a book of advice for aspiring engineers in 1934. He described the objectives of engineering education in terms that clearly assume a genetic basis for success:

> Education means cultivating, developing and training the native faculties in their function and coordination. . . . College, with its teachers, laboratories, and libraries constitutes merely an environment favorable to such development. . . . The student's possibilities of education are limited by the inherited powers, just as in physical development his final stature is limited by native physique regardless of any efforts toward his physical training.[104]

A peculiar tension lies under the surface of Williams's pronouncement, and in the entire consideration of inherited characteristics as the basis of good engineering. Are the skills required for testing and inspection teachable, or not? Seemingly, in instructors' genetic categorizations of students lay an argument for the scholars' own irrelevance. Charles Dudley anticipated this very issue: "Experimenters are born, not made. . . . [But] if so much depends upon cast and habit of mind, what can the schools do in the way of training and furnishing mental equipment to produce successful testing engineers? We answer much, in every way."[105] Schools, he pointed out, could not make "a successful engineer out of a numbskull," but they could certainly succeed with good raw material. John Price Jackson expressed a similar confidence in the university when he noted that "lazy" or "unambitious" students might be restored to "a true understanding of the meaning of their college course and to an enthusiastic interest in their technical studies" by being brought into proximity to "more active and right-minded classmates."[106]

Whether these judgments correlated primarily with a student's economic, religious, or ethnic background is a difficult question, but we can get one step closer to understanding how they functioned as mechanisms of social control by considering their use in a frankly "nonelite" setting. Between 1900 and 1910, directors of the Lewis Institute of Chicago often noted that their charter was to provide practical training to "deserving" people of Cook County who might not otherwise have a route to gainful employment. The word *deserving,* as Michael Katz has shown, carries a sense of the variable sources of poverty perceived in modern America, a perception whereby some underprivileged persons were thought "worthy" of corrective aid and others not. For the Lewis Institute a lack of money was not meant to be of itself an impediment to self-improvement; rather, some other moral or social element of each student was to determine his occupational destiny.[107] Taken together, these formulations show that for engineering instructors of this period, university attendance was a necessary component of engineering training but not a sufficient one. This strategic combination of innate and acquired abilities for engineers gave both self-justification to the instructors and a reasoned, if narrowly conceived, basis to their elevation of white male students to professional eligibility.

Conclusion

As part of his research for the Carnegie Foundation, Charles Mann polled professional engineers on the question of which qualifications they deemed important for success. After a series of inquiries to some fifty-four hundred practicing engineers, Mann summarized his findings, listing "characteristics" according to what percentage of respondents rated them as most important:

Character, integrity, responsibility, resourcefulness, initiative	24.0
Judgment, common sense, scientific attitude, perspective	19.5
Efficiency, thoroughness, accuracy, industry	16.5
Understanding of men, executive ability	15.0
Techniques of practice and business	10.0[108]

The historical meaning of this survey is twofold. Fewer than half as many engineers found "techniques of practice and business" to be most important to a successful career as found "character" and its correlates to be most important. One in four engineers seems to have believed that character outweighed knowledge in the conduct of a successful technical career. These men may have found matters of personality more helpful than their technical skills in day-to-day work—a reasonable and provocative possibility that would best be tested against personal accounts.

But we can also use Mann's tabulation to get a more general sense of how observers—Mann, the Carnegie Foundation, the journals that published these findings, the men who faithfully answered the questionnaire—saw the field of engineering. The choices offered to respondents show a preponderance of nontechnical categories. Whatever the relative weight the survey's sponsors and respondents gave such categories, in either pedagogical theory or professional experience, the explicitness of these features of engineering practice reveals that they had a utility for people engaging in technical work. The design of the survey and its findings, whether or not provided in good faith, give to engineering a high ethical and moral profile.

The presentation of character, religious practice, gender, race, and even voice and hair color as matters pertinent to technical operations fits with the subjective nature of technical work outlined within engineering coursework. That coursework, as we have seen, made the educators' invocation of personal qualities seem a reasonable feature of engineering training. In turn, the rhetoric of educators and professional engineers on the importance of morality to professional practice bolstered classroom invocations of intuition and judgment. Quantification was by no means incompatible with such qualitative practices; in the minds of educators and engineers the two approaches supported one another and together comprised effective materials testing in early-twentieth-century construction.

That educators shaped this service to industry with a great sensitivity to the public image of engineering is obvious. Prominent engineering instructors such as Edwin Marburg, Arthur Talbot, and Anson Marston, and many of those who joined the Society for the Promotion of Engineering Education, clearly saw part of their work as being to craft an attractive identity for high-level (and costly) technical work in the commercial world. But we need not reduce their concern with high moral and cultural standing to a merely promotional project. Their desire to secure a wide and lasting commercial presence for their discipline—that is, certain and high-status employment for their students; industrial projects for themselves and their universities—arose also from the idea that only certain kinds of behaviors yielded effective quality control for industry. To carry outward from the academy that particular array of behaviors—both technical and personal, for the two were inseparable—the academics worked closely with their industrial colleagues to produce written regulations for materials testing. The published standards and specifications that joined young testing engineers in the commercial sphere, disseminating new knowledge and procedures about concrete to the building trades, are our next subject.

CHAPTER TWO

Science on Site
The Field-Testing and Regulation of Concrete Construction

Rules cannot produce or supersede judgment; on the contrary, judgment should control the interpretation and application of rules.
—*American Concrete Institute, 1917*

We will now follow the growing body of knowledge about concrete outward from its academic home to its application on the early-twentieth-century construction site. In some ways the scene shifts dramatically, from the orderly classrooms and laboratories of the university to the grit and commotion of the commercial building site. Concrete in the first decades of the century was used primarily on very large building projects, such as factories, warehouses, dams, and shipyards, and most of these structures were erected under the rushed conditions of modern profit-based enterprise. The graduates of engineering programs thus brought their meticulous and cerebral work out into places of almost frantic activity. Crowds of laborers bearing shovels and brooms moved among huge steam-powered cranes. Systems of chutes and carts shunted raw materials—sand, gravel, reinforcing rods, wet concrete—from stockpiles to the rising building.

Surrounded by these constant movements of men and machines, the college-trained technicians scrutinized raw materials and portions of the finished structure, testing and inspecting hundreds if not thousands of specimens on any given job. But the pressure and pace of the concrete construction site were not deterrents to the testers' work; those circumstances were the very reasons that their work existed. After all, from its inception, proponents of concrete testing had sought to combine the formal features of systematized science and the technical adaptability required of a busy, cost-conscious commercial service. As promised by the experts who had trained them, the testing engineers' on-site work prevented waste and assured the integrity of structures, bringing science-based uniformity to the handling of concrete on a massive scale.

It is no surprise, therefore, that testers' knowledge and techniques were able to stand up to the rough and ready conditions of commercial application. What does require exploration is the fact that techniques of concrete testing did not disseminate *further,* into more efficient and cost-effective applications in the hands of more workers. In crafting scientific techniques of concrete testing as a set of teachable skills, university instructors had codified and instrumentalized many features of their discipline for export to the building industry. Once they saw the utility of science-based testing and inspection, why did the owners and managers of construction firms not press for more mechanization, more kits, more simplified field procedures that could be put into the hands of lower-paid employees? Why not demand cheaper and more thorough translations of science into a tool of routine production or some other subversion of the trained testers' specialist identity? We have seen that instructors in the flourishing field of materials testing sought to link quality control of concrete to particular claimants: the men who had passed through the university programs. We need to ask how they succeeded. In the very fact of its dissemination—its translation into techniques and instruments for use in the field—science-based testing threatened to self-destruct as an expertise under close jurisdictional control. What's more, most other features of concrete construction—almost all physical aspects of its handling, in fact—had been subject to the de-skilling and divisions of labor popular among modern business managers. Much of the material's commercial appeal derived from its departure from traditional craft-based building techniques. How did the field-testing and regulation of concrete achieve its autonomy and solidity as a field of expertise in the face of such powerful trends?

As suggested earlier, the high intellectual regard in which industry held science is not the answer here but rather the real question. This group of testing engineers, far from the laboratory or construction company front office, stood up to their knees in dust and wet concrete yet somehow held on to a modicum of occupational prestige denied to all those around them on the site. Their work certainly kept buildings standing, but so did that of bricklayers and carpenters. Among hands-on practitioners only the materials specialists induced building firms to recognize a particularly high social and economic valuation of their service to industry—as that usually reserved for the work of managers. Crucially, the technicians were not managing other employees on the construction site in any direct sense. Their findings about materials and construction operations may have determined what laborers did once those finding reached site managers, but the work of testing itself was physical, not supervisory. Its elevation by the building industry to the status of expertise was not a foregone conclusion.

The success of materials experts in commercial construction rested on two initiatives. On the one hand, the specialists shaped science-based techniques of quality control for concrete as a labor process, as a particular set of work rules and job classifications. Recommendations for testing and inspection, and the promotional rhetoric surrounding such protocols, all pointed to a particular division of labor on the concrete construction site, with the main work of quality control falling to men of particular credentials. In this way materials experts reiterated the prevailing hierarchical organization of labor on the construction site, facilitating the direction of large numbers of semitrained workers by a few highly trained individuals. The materials experts offered their clientele a means of de-skilling other workers while avoiding such a diminishment of their *own* occupation. At the same time, the materials experts became in effect brokers of technical knowledge. Like most large-scale business enterprises of the day, construction was enveloped in a complex of competitive relations, between suppliers and buyers of services and between competing suppliers of the same product or service. Reputational and legal contestation was a constant feature of these relations. Materials experts refined their services with great sensitivity to the pressures faced by their clients; scientific techniques for testing and inspecting concrete were themselves commodities for sale and exchange in the world of commercial construction. In short, this was science on the side of commerce, helping control the organizational and competitive pressures felt by the twentieth-century building firm. Both of these accommodations to industry were brought about by powerful "literary technologies" used to disseminate new knowledge and techniques for construction: published standards and specifications for concrete.[1]

These written protocols packaged scientific knowledge for routine use by industry. Materials standards outlined optimal performance characteristics of an industrial material such as cement or concrete. Specifications, often incorporating standards, laid out the specific design or material characteristics of a given product or project—in this case, a building or building type. Both recorded experts' conceptions of "best practice" for conveyance to dispersed sites of technical activity, usually as the basis for estimates or contracts exchanged between the buyers and sellers of a technical service.

The parentage of standards and specifications was academic and commercial. Increasingly after 1900, scientific specialists in the strength of materials, employed as academics or consultants, contributed their talents to standard-writing committees of the American Society for Testing Materials or the American Society of Civil Engineers. They were joined on these committees by representatives of industry and by technical experts in related fields; in the case of cement and concrete, these experts included building designers and sanitary or hydraulic engineers. The committees assembled collections of standards and sample specifications, submitted them to review by their parent organizations' membership, and after approval published the protocols for use in the commercial world.

The joint origins of these instruments in academia and commerce immediately suggest some utility for them beyond the improvement of industrial processes. This chapter focuses on the first function of standards and specifications mentioned above: that of carrying forward the hierarchical occupational vision of the university engineering instructors to associate testing and inspection jobs with highly trained practitioners. Managers and owners of building firms shared this vision, but the movement of bodies alone from classroom to job site could not assure that testing would remain an elite specialty in the competitive and profit-driven world of commercial construction. The written instruments encouraged lines between "inspectors" and "inspected" to follow those of educational background (and thereby of class and gender), as outlined in chapter 1. (The second social function of science-based quality control for concrete, the brokerage of scientific knowledge to control competitive pressures among commercial concerns, will be examined in chapter 3.)

In rendering complex productive tasks more predictable and bringing about a close control of material and human behaviors, standards and specifications for concrete had much in common with the procedural guidelines emerging for manufacturing and office management in this period. To bring to the foreground the occupational power of standards and specifications in the modern workplace, we will relegate to secondary status the fact that these regulations also worked in the technical sense. Concrete buildings erected with modern science-based quality-control methods rarely collapsed, and certainly the new rules contributed to the popularity and steadily decreasing cost of building with concrete after about 1910. But we will focus here on the significance of concrete standards and specifications as new administrative and supervisory devices for the day-to-day business of construction. In this capacity the written regulations exemplified and promoted contemporary trends in industrial organization.[2]

As ideas of systematic and scientific management gained credibility within many manufacturing contexts, business owners mechanized and subdivided production tasks. Decisions about product direction or design were removed from the shop floor or, in the case of construction, the work site. In construction, many traditional craft-based building practices were gradually being replaced by routinized procedures. The diversified skills of carpenters or masons trained in the apprentice system held little attraction for most modern construction firm owners, and divisions between the roles of construction workers and managers—who could be contractors or firm owners—began to evolve accordingly. Conflicts about site operation arose, and the higher reaches of the building trades expressed antilabor sentiments of new intensity, all of which encouraged managers to turn the responsibilities held by building site workers over to a few supervisory personnel. The materials experts' prescriptions for the field control of concrete fit with these managerial goals by reducing the autonomy of lower- and lower-middle-echelon workers.[3]

Again, what is intriguing is not that lower-paid laborers had little responsibility on the work site but how the university-bred skills of the testing engineers overrode employers' desire for economies. To understand this delegation, we need to return to the idea, expressed in engineering pedagogy of the period, that science-based testing for concrete embodied a combination of relatively systematized assessment techniques and a substantial body of much less objective operations, such as the use of judgment and intuition by the tester. This fundamentally doubled nature of concrete testing was established in the design of coursework in materials testing and also formed the essence of standards and specifications after 1900. In these written instruments we find many invocations of distinctly unroutinized practice.

At first glance, many standards and specifications of this era, like the teaching methods of engineering instructors, seem distinctly unscientific or even sloppy in their retention of vague or qualitative terminology. For example, a 1911 specification for reinforced concrete notes, "The test load applied shall not exceed that which will cause a total stress in the reinforcement of 3/4s of its strength at the yield point"—a fairly straightforward quantitative point. But the specification also requires that "the entire work shall be constructed in a substantial and workmanlike manner."[4] Despite its relative generality, the second phrase was as carefully chosen as the first; its lack of specificity helped link quality-control practices to university-trained men. The wording of standards and specifications was not poorly conceived but was rather the artifact of a rationality we do not normally seek in technical bodies of knowledge: the successful pursuit of occupational and social goals.

History of Standards and Specifications for Building

Standards and specifications are tools of communication for the execution of technical work. Through the first decades of the twentieth century they aided the work of building at several junctures, and we should begin by sketching the role of these written instruments in the construction of a typical well-capitalized building of the period.

First, an architect or engineer would design a structure and generate from the design a series of detailed drawings and written specifications to make clear all elements of the project. Dimensions, forms, materials, and performance for all parts of a structure required careful expression in these documents to communicate the ideas of designer to the project's "owners" (those commissioning a building). Additionally, a building's designer or other member of the designing firm would use the drawings and specifications to create financial estimates, beginning the process of predicting and controlling costs for this work. Then the designing firm, if it was also responsible for the erection of the building, issued requests for bids and ultimately contracts—all crafted to communicate features of the design to the contractors who would undertake the actual construction. In the case of larger factories and other concrete structures after 1900, engineering firms, rather than architectural firms, usually assumed this supervisory role.

Those bidding to work on the building—such as electrical engineering firms, plumbing firms, cement producers, or more general building companies—incorporated additional specifications into any plans they produced for the project. These they submitted to the engineering firm for acceptance. Finally, and most significant for this telling, standards and specifications had a further life on the construction site itself, directing the testing and inspection work undertaken there to assure the quality of materials and structural elements. Testers and inspectors used the specifications issued by the engineering firm in charge of the project to make sure contractors and suppliers conformed to expectations.[5] Contractors and suppliers sometimes performed tests to confirm their own performance and that of subcontractors. Thus, standards and specifications, embedded in plans, bid, and contracts, carried technical ideas and commitments from group to group over the course of a building project. Amended and adjusted to reflect the material and economic conditions faced by each group, they were instruments of great pecuniary and administrative power.

As the word *standard* suggests, those writing specifications for a particular building or building type did not create their instructions from scratch. They turned to the sample specifications increasingly available after 1900, taking advantage of the experience of others who had faced similar logistical challenges. This body of published specifications for construction came readily to hand. In early commercial applications of scientific quality control, such as those of the metal industry or very large manufacturers of the mid-nineteenth century, a company might have issued a manual of inspection criteria and practice for its own personnel. Collections of standards and specifications, published by scientifically oriented trade associations, increasingly replaced or augmented these manuals after 1900. The earliest efforts to coordinate such evaluative techniques to reflect industrywide conditions may have been the Franklin Institute's Boiler Codes of the 1830s (a federally funded project), followed by the creation of comprehensive rules for metallurgical testing by railroads and steel industries at midcentury. But by the last decades of the century the electrical and chemical engineering fields had made such projects their specialty and carried them to a wide range of industrial settings, from mining to power generation to food processing.[6] In the years following 1900, civil engineers addressed the difficulties of erecting buildings, dams, bridges, and the many utilitarian structures associated with modern manufacturing with their own techniques of material assessment and analysis.

American business owners by this time had come to realize the benefits of intercompany standardization. They supported the creation of standards and specifications through new committees of the ASCE, the American Railway Engineering Society, and increasingly the ASTM. The ASTM had been founded in 1898 when a group of Philadelphia businessmen and engineers formed a freestanding section of the International Association for Testing Materials. (The IATM itself had been started in 1895 by Europeans and Americans who had worked together informally on problems of materials testing since about 1880.)[7] While a handful of observers felt that sharing proprietary information of this kind worked against what they considered to be healthy capitalist competition, many industrialists realized that pooling certain types of technical knowledge saved duplication of investigative effort and satisfied consumer demands for compatible parts and materials. In addition, the idea of quality control may imply *upward* limits of minimally acceptable performance for a raw material or finished product. Industry standards can moderate the general level of precision and quality available to, and reasonably demanded by, consumers and increase producers' economies in this way as well.[8]

A parallel initiative on the part of the federal government gave rise in 1902 to the National Bureau of Standards (NBS, now the National Institute of Standards and Technology). This body grew out of the much smaller Office of Weights and Measures of the Coast and Geodetic Survey and served to centralize most testing efforts of the United States Army, the Department of Agriculture, and later the United States Geological Survey. Initially created to standardize a wide array of materials and goods purchased by the federal government, the NBS steadily extended its purview to address the many logistical problems associated with massive growth in American industry, including chemical, electrical, and structural obstacles to safety or efficiency. Its efforts included basic research in the behavior of materials, the development and testing of instruments, and tests on selected commercial specimens. The bureau began publishing specifications in 1912 with a circular on Portland cement; by 1928 it had prepared several hundred specifications for products that included mucilage, fire brick, rubber tires, shoe leather, and asphalt.[9] But the bureau remained dependent on congressional appropriations, and while helpful to industry as a source of technical knowledge, it could not respond fully to the demands of the diversifying industrial sector. Nor was a government bureau likely to accommodate closely the economic concerns of the private sector by tying its researches to minor technical innovations or fluctuations in materials costs.[10] Moreover, many business owners, fearing federal regulation, might not have wanted to encourage government activities of this type. Other government-sponsored investigations of materials, such as those conducted by the Army Corps of Engineers on building materials, functioned only as secondary resources for commercial concerns.[11] The ASTM coordinated research conducted by many smaller organizations and universities, and for many American businesses after 1900 it was this body that increasingly came to represent a private, proactive, centralized source for standards of best practice.[12]

The ASTM did not maintain its own laboratories but instead gathered data and promoted research at academic and commercial sites, collating the results. No product in which American industry had an interest—from trolley wire to paraffin, rubber cement to plywood—escaped the society's scrutiny. It devoted its attention to a full range of raw and finished industrial materials—steel rails, wrought iron, brass and bronze, wires and cables, pigments, timber, rubber, and textiles—and those materials associated with the emerging technologies of concrete construction: reinforcing bars, cements, aggregates, and waterproofing compounds. Each was dealt with by a specialized committee that was, as all involved seem to have considered most fair, constituted of both materials "producers" and "nonproducers."[13] The former category included owners and employees of manufacturing, building, engineering, and other businesses. The latter category was filled largely by scientists and engineers employed as university professors, with some government- and self-employed participants, along with potential purchasers of the material or product under scrutiny. Once a set of standards had been written, reviewed by the general ASTM membership, and published, that membership—a much larger audience of business and government officials—incorporated the directives into contracts, company manuals, city safety codes, and other documents intended to regulate the material features of commerce. Contemporary textbooks recommended inclusion of ASTM standards in all types of contracts. By the 1920s an elaborate legal web had been woven around the standards, directing liability through a careful attribution of responsibility based on these standardizing prescriptions.[14] For large-scale commercial concrete construction projects, engineering firms spliced appropriate portions of published standards and specifications into masses of detailed documents, controlling to a remarkable extent the complicated technical, commercial, and legal operations of the modern building project.

With a sense of their institutional origins in mind, we can now consider the patterns of work organization, the exact delegations of skill and authority, that the standards and specifications brought to the modern workplace. In converting their knowledge to written form, particularly to series of simplified instructions, the authors of modern standards and specifications intended from the start to put the application of their knowledge in hands other than their own. Neither the technical nor the business-based members of ASTM committees intended to use the codes for their own daily work. The essence of such codes, after all, was to make science an affordable tool for producers, builders, and other bodies, such as city governments, involved in the production or use of modern materials. But the wholesale reduction of science-based quality-control procedures to minimally skilled labor would have undermined both the authors' own claims to expertise and those of anyone employed at such work. Standards and specifications for concrete construction offered a middle way. This would be an organization of quality-control work on the building site that rendered this service affordable—that is, less expensive than the services of expert consultants would have been—but that established it securely as a set of tasks to be performed by relatively skilled personnel, *not* laborers.

We need to understand first which physical features mattered to those building with concrete—the concerns and problems encountered by builders that science might resolve. Some of the difficulties of using concrete to which science contributed aid involved matters of design. The optimal form of structural elements and the arrangement of reinforcement within concrete elements garnered the attention of many civil engineers in business and academic positions. Issues of interest to materials specialists centered on the activities of the concrete construction site itself, where raw materials combined into finished building. Here concerns about costs, the technical exigencies of concrete construction, and ideologies of work organization also came together to create a particular system of opportunities and risks on the building site. We can then look at the content of standards and specifications that allowed technical experts to enact for industry a mutually advantageous organization of work on the construction site.

Defining the Nature and Safe Use of Concrete

The concerns that builders and technical experts had about concrete involved a disjunction between its ideal and real characters. These concerns focused, in 1900 as today, on the idea that concrete *flows*: it is a plastic medium that achieves its solidity and strength some time after it has been put in place. Ideally the use of concrete offered turn-of-the-century builders an uninterrupted production sequence moving as smoothly as any assembly line of the day, generating buildings rather than cars or sewing machines or canned peaches. In actuality that flow might easily result not in a sturdy, speedily erected structure but in a weak and costly mess. Concrete is vulnerable at many points to small and large deviations in its handling.

To achieve the steady flow of wet concrete around a building site to waiting forms, builders based concrete construction on systems of pipes, chutes, buckets, and conveyors that distributed wet concrete from a central mixing area. One portion of a building could be poured while another hardened, each floor standing on completed portions of the one below. But obstacles to this efficiency abounded. Concrete was said to be a difficult material because it acts as a deadweight when wet: it cannot support its own weight until hardened. Therefore it is not only cumbersome but requires that its handlers be able to judge when it has hardened. Furthermore, concrete building frames cannot function reliably unless the junctions between beams, columns, and girders are contiguous. Those responsible for building with concrete must have an understanding of what constitutes a true join between elements poured at different times.[15]

Possibly the point of greatest concern for those directing the field use of concrete after 1900 was the impact that the ratio of water to cement could have on the ultimate strength of a concrete mixture. The interpretation of the optimal water/cement ratio changed as the use of concrete became more popular as a commercial medium. Too little or too much moisture in a concrete blend will undermine the chemical processes by which a beam, slab, or column "cures," or hardens. Laboratory investigations might readily resolve such technical questions as how much water, cement, gravel, and sand make the "best" concrete in some narrowly defined way, but the conditions of commercial construction impinge heavily on such questions. The nineteenth-century French engineer Louis-Joseph Vicat wrote that the "ancients insisted" on mixing concrete with virtually no water: "Good mortar, they said, ought to be tempered only with the sweat of the mason."[16] This is obviously an exaggeration, but mixtures did remain relatively dry until about 1900, the start of the "wet-mix era." The shift occurred because wetter mixtures flow into place more quickly, fill forms more readily than dry mixtures and require less tamping, and thus could save time and money on the larger projects of the post-1900 period.[17] Wetter mixes also save money for builders by reducing the quantity of dry materials consumed (water costing less than sand, gravel, and especially cement).

But difficulties arose for advocates of wet mixes because as attractive as such blends were in terms of handling, even slightly too much water could be detrimental to the durability of the concrete. Both building buyers and the engineering firms that designed concrete buildings worried that without strict regulation, contractors erecting concrete buildings would overdilute their mixes and produce weak structures. It is worth noting that this was long before the advent of the familiar rotating "ready-mix" truck that now carries premixed wet concrete from a central plant to the building site. Until the 1930s all concrete was mixed on the site, thus preventing truly centralized control of mixes. But engineers and building buyers, like the contractors they employed, also wished to hold down costs in terms of cement and man-hours; they too wanted to achieve as wet a concrete mix as possible, within reason. The tolerance of concrete for variations in wetness became a topic of much debate as testing specialists, designing engineers, cement producers, and builders attempted to find the fastest means of erecting sturdy structures.[18]

Thus, experts recognized that in a number of ways the expediency offered by concrete as a pourable medium could be offset by its need for precise handling. The development of concrete from this point forward depended on the fact that concrete experts and cement producers approached these technical difficulties as subjects for treatment within their preconceived system of occupational relations. The operations of the modern engineering firm will be elaborated in chapter 4, but we can note here that the continuous nature of concrete construction can be contrasted readily to the "unit-based" operations of brick or stone masonry and carpentry, and that the pursuit of flow brought not only a faster means of building but also the use of a new labor pool. Construction firms used some conventionally skilled operators on concrete projects, such as carpenters to build custom-designed forms and metalworkers to fabricate unusual pieces of reinforcement, but most of the work could be handled by workers with little training who moved raw materials and wet concrete around the building site. Most tasks were reduced to simplified series of repeated actions, determined to the most minute detail before the project was begun.[19] This allowed construction companies to pay lower wages to individual employees, even if overall numbers of employees were not reduced, and also to avoid using union labor if they so chose since so many workers on a concrete site could be hired off the street (and be easily replaced if found to be unsatisfactory).[20] With these broad managerial precepts in mind, concrete experts and their industrial colleagues generated written technical recommendations carefully conceived to ascribe blame, direct responsibility, and generally differentiate tasks of construction according to a hierarchy of—in descending order of authority—scientific or engineering experts, materials manufacturers and construction business management, and labor.

This quality-control system of standards and specifications had at its foundation a rhetorical construction of blame for material problems on concrete construction sites. Engineers, cement experts, and producers began by publicly blaming concrete malfunctions on operations at the site, instead of on the quality of the cement, their own prior recommendations for mixing the concrete, or building design. Discussing the dangers of using too much water in concrete, experts and cement producers associated the risk with two very specific field practices: the propensity of workmen to dilute concrete mixtures that had become dry because of delays on the work site, and their tendency, when so directed by the contractors for whom they worked, to stretch cement with water because water was the cheaper ingredient.[21] Similarly, a speaker at the 1912 meeting of the American Concrete Institute offered an extensive list of potential errors by concrete construction workers: too early removal of forms; careless and unwise placing of reinforcement; inadequate provision for expansion and contraction because of changes in temperature; the use of weak and leaky forms; and inadequate tamping of wet concrete.[22] All of these formulations directed blame for problematic structures away from the creators of raw materials or building designers and towards fieldworkers.

These ascriptions helped suppliers of goods and services defend their offerings, and it is not surprising that editors of *Cement Age,* for example, applauded makers of cement. But beyond the level of promotional rhetoric, the remedies suggested for these problems reinforced existing occupational hierarchies and systems of economic opportunity. Cement experts, cement producers, and building industry managers militated for practices that would keep decisions about fieldwork in the hands of university-trained engineers or, at the very least, above the level of artisans or laborers in the building trades. A more direct solution might have been to train the people who actually handled concrete on the construction site to judge its proper treatment. This would have been expensive,[23] but more important, it would have threatened the skill-based distinction between the testing engineers, business owners, and managers who dictated the safe use of concrete, on the one hand, and those who labored with the material at far less pay and with far less occupational mobility, on the other.[24]

Instead, experts and industry managers sought to achieve field control of concrete while systematically removing judgments about materials and designs from the province of field laborers. Far from presenting a conundrum, this paradoxical situation offered a professional opportunity to on-site technicians: a means for managers to allocate responsibility for good concrete buildings. To probe this allocation, we need to highlight one additional characteristic of specifications. Even where the various actors contributing to the quality control of concrete agreed upon optimal technical practice, the authors of quality-control protocols could choose from among many different ways of achieving safe and efficient construction. Each carried a particular delegation of responsibility in the commercial sphere. Whether addressing the width of a window sill, the depth of a floor slab, or the sturdiness of an entire foundation, a specification could attach the fulfillment of that task to a given participant on the project.

An illustration of how this worked is provided by Daniel Mead, an engineering instructor at the University of Wisconsin turned private consultant, who in 1916 wrote a lengthy textbook on the authorship and use of specifications, intended for engineering students preparing for construction industry jobs. He offered a concise listing of different types of specifications, which I have annotated here to show the associated distribution of responsibilities each would institute. According to Mead, specifications could bring about a quality product:

> a) By delineating composition or properties, and manipulation or workmanship *[this gives the supplier responsibility for providing materials of a certain character and gives the contractor the job of making sure that certain operations are performed on these materials]*
>
> b) By inspection *[demanding assessment of contractors' activities or materials by a qualified inspector, employed by the engineering firm in charge of the building's erection]*
>
> c) By guarantees *[deflecting responsibility onto the contractor or supplier without specifying further the nature of the work itself]*
>
> d) By tests *[akin to inspection in delegating responsibility to a technical expert chosen by the engineering firm rather than relying on the contractors' adherence to any descriptive details provided in the specifications themselves]*
>
> e) By selecting materials or suppliers used successfully under similar conditions *[here the engineering firm again takes responsibility by directing contractors towards acceptable sources]*
>
> f) By requiring the equivalent of a selected standard *[again, making the engineering firm responsible for identifying acceptable standards, rather than providing a list of detailed procedures]*.[25]

All of these approaches found a place in the plans and contracts of engineering firms that erected concrete buildings. Some controlled the "lateral" relationships in the building industry between engineering firms and the contractors and suppliers in their employ. Others among the approaches that Mead lists mediated the "vertical" relationships between more and less qualified workers on a building site, following divisions along lines of skill and workplace authority. Here we focus on testing and inspection by representatives of engineering firms, procedures that were in great enough demand that young men trained in these areas in college engineering programs found a steady source of employment in such positions. The content of specifications—the descriptions of testing and inspection procedures they provided to engineering firms—were instrumental in achieving that occupational pattern.

Testing in the Field: A Search for Uniformity

Materials experts, cement producers, and engineering firms began their concerted effort to bring testing into the routine business of concrete construction around 1900. Scientific procedures for construction were sometimes introduced through the provision of elaborate field laboratories that could perform chemical and microscopic analyses of materials, but only on the largest projects.[26] Far more frequently, engineering firms performed tests of cement and concrete on the building itself or in simple shedlike shelters. Different tests determined the quality of cement, the character of a concrete mixture, or a structural element's final strength. The tests centered on characteristics of dry cement, samples of cement mixed only with water (called "neat" specimens), wet concrete mixtures, or completed elements of the building. For dry cements, experts considered specific gravity and fineness to be reliable indicators of quality: the former could be assessed with a simple calibrated glass apparatus, the latter with a series of graduated sieves (fig. 2.1). After 1900 most construction sites for buildings above the size of family homes were under the immediate supervision of a general contractor and the ultimate control supervision of a superintending engineer. A staff of assistant engineers or inspectors conducted tests according to their orders.[27]

To facilitate this on-the-spot testing, some simplified testing technologies appeared soon after 1900, and we have seen these in use in the university laboratories in which testing personnel had received their training.

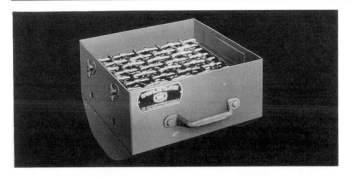

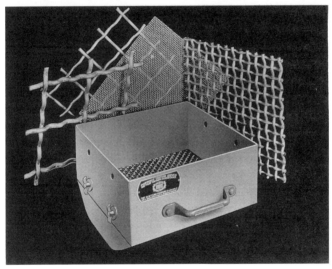

FIGURE 2.1. Portable sieve set produced by the W. S. Tyler Company for measuring gravel used in concrete in the early 1930s. Such kits permitted concrete users to comply with standard specifications widely incorporated into city building codes and commercial contracts, and they functioned as both legal and supervisory instruments for construction firms. *Iowa State University Archives, Gilkey Papers*

Kits and portable instruments were produced for testing the composition of cement or the soundness of concrete while a building was under construction and were marketed as time- and labor-saving innovations (fig. 2.2). These devices were often accompanied by preprinted forms on which records of test results could be entered to assure regular testing and easy comparison of results over time.[28] The "slump test" for judging concrete's consistency, still in use today, also emerged at this time. A metal cone about

FIGURE 2.2. Portable apparatus for assessing the composition of cements and concrete mixes, produced by the Humboldt Manufacturing Company of Chicago in the late 1910s and intended for field use. *Iowa State University Archives, Marston Papers*

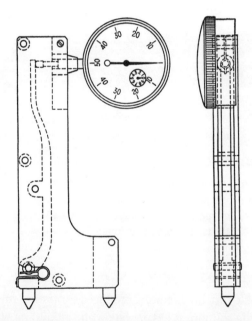

FIGURE 2.3. Portable "2-inch" strain gauge for measuring the behavior of concrete during and after hardening. This gauge and others were designed by Professor H. C. Berry of the University of Pennsylvania, who from the 1910s onward manufactured his gauges and marketed them to academic and commercial purchasers with great success. *From Herbert Gilkey, Glenn Murphy, and Elmer O. Bergman, Materials Testing (New York: McGraw-Hill, 1941)*

12 inches high would be filled with wet concrete, then inverted to leave the concrete standing on its own. The degree to which it subsided gave an indication of its wetness. Assessment of time of set, tensile strength, and constancy of volume for wet cements were only slightly more complicated. Most rested on the production of sample cylinders or briquettes (small hourglass-shaped pieces) by pouring mixtures into molds, letting them set, and then subjecting the resulting pieces to inspection by eye or instrument after seven, fourteen, twenty-eight, or ninety days. Briquettes might be created either under conditions identical to those of the building under construction or under "exaggerated" conditions—through exposure to cold, heat, or steam—that could provide a reliable indication of the material's ultimate performance.[29]

Speed and ease of use were in many ways the defining criteria for the design of these tests. For example, by using "gang molds," advertised widely after 1910, inspectors could quickly produce duplicate briquettes of the same mixture and easily conduct corroborating or comparative tests. Tests of complete beams, slabs, or columns, while in some ways demanding the greatest theoretical knowledge of design, achieved the same efficiency of execution. H. C. Berry, a professor at the University of Pennsylvania closely involved in the development of ASTM specifications for cement, patented during his career portable strain gauges of increasing durability and simplicity (fig. 2.3).[30] Simplified portable field models of the compression- and tensile-testing machines used in materials research laboratories also reached the market in this period.[31]

However simple and quick these testing procedures may have been, their reliability was dependent on the preparation and handling of the specimens under test. These were processes that could vary considerably. Materials experts scrutinized field tests of cement and concrete and found numerous factors that could skew results even among tests made by a single operator. These factors included irregularities in raw ingredients, different durations of mixing, the density with which wet cement was packed into a mold, defects in molds, the ambient temperature under which a test was conducted, and the manner in which a testing instrument was operated. More complicated tests were subject to even further irregularities, such as the improper placement of a specimen in a machine or variation in the rates at which test loads were applied. These variables meant that it was often difficult, if not impossible, to judge a specimen against known standards or to compare specimens in the course of a construction project.

Several approaches presented themselves as ways to overcome this variability. First, multiple tests could have been made in any situation and then averaged. This would have been costly and time-consuming, however, and would have defeated the objective of obtaining immediate results in the field. Also, the averaging of discrepant test results was not considered to be a safe procedure because it could obscure real weaknesses in specimens.[32] In 1902 one engineer suggested a "grass-roots" approach to the problem, whereby the many users of cement and concrete would voluntarily tabulate their test results to obtain a centralized body of data against which routine findings could be measured. This approach would have been difficult to coordinate, especially since raw materials varied by locale. As we will see later, it was probably unrealistic to expect many competing cement producers or consumers to share the fruits of their own research. In 1906 Robert Lesley, the cement manufacturer and publisher, advocated the creation of a centralized private laboratory, to be a "supreme court of testing in the hands of those representing the highest advance in the art." This suggestion also failed to address the problem of immediate on-site testing.[33]

Instead, the scientists and engineers who dominated professional societies after 1900 pushed for the development of uniform testing methods, to be disseminated as portions of published specifications. These instructions would detail every step in a test specimen's preparation and handling and thereby, they believed, forestall the many irregularities afflicting field-testing of cement and concrete. It was a solution embraced by cement producers and reinforced-concrete building firms, and one that reflected a number of technical and professional agendas.

Defining Uniformity

Uniform testing for cement and concrete was not a new idea. As early as 1879 the American Society of Civil Engineers had formed a committee to devise a uniform system for tests of cement. (Tests of concrete were not in common use at this time, and most tests of cement were intended to ascertain the quality of a cement before its incorporation into concrete.) The committee's final report, issued in 1885, recommended a series of fineness, cracking, and tensile tests for any application of cement. The report specified procedures for mixing materials (including the time and manner of mixing) and for filling molds (cement was to be placed with a trowel rather than rammed). It also specified the rate at which stresses should be applied in tensile tests.[34]

By 1891, French governmental engineers had also issued a set of instructions intended to regularize cement testing. These were shared at international engineering congresses over the next decade.[35] Along with new, more precisely graduated instruments, the American and French instructions represented a clear improvement over earlier conventions for dictating test procedures. However, by 1897 the U.S. cement industry and many engineers who worked with concrete were complaining about the elasticity of existing regulations. In 1898 the ASCE formed a new committee to examine the problem of "manipulating" tests of cement, as materials experts termed the operations of specimen preparation and handling. The organization defined the regularization of manipulation as a problem distinct from that of deciding which tests to perform on cement or concrete. This distinction indicates the importance cement and concrete specialists gave to manipulation, and their understanding of it as something subject to control from "above" through their own initiative rather than something to be left to the people actually conducting tests.[36]

The ASCE's Committee on Uniform Methods of Test began its study of how great the variety of testing procedures might really be by surveying seventy-five members of the society who used cement in their work. On the basis of fifty replies, the committee members declared a "decided lack" of uniformity in testing, "nearly every engineer having methods of his own origination."[37] They responded to the problem by refining instructions for many aspects of the testing process. For fineness tests conducted with sieves, the committee added the amount of time a sieve should be shaken (1 minute) and the rate at which it should be shaken (200 strokes per minute). They cited desirable temperatures for materials under test and for testing rooms ("as near to 21 degrees centigrade as it is practicable to maintain it"). They described how materials should be mixed: stirred with a trowel and then "vigorously kneaded with the hands for an additional 1½ minutes, the process being similar to that used in kneading dough. A sand-glass provides a convenient guide for time of kneading."[38]

These instructions became the basis for recommendations on cement use that were issued by the American Society for Testing and Materials in 1904 and that found a wide audience through the ASTM's endorsement. The ASTM concerned itself with both the regularization of methods of testing and the standardization of materials and products, considering the two tasks to be inseparable.[39]

Prior to adopting the ASCE's instructions for testing cement, the ASTM asked thirty laboratories to try the ASCE's methods of test. The laboratories gave a favorable report, and the ASTM's committee of thirty-two cement consumers, manufacturers, and professional inspectors recommended them to the society. After the society's membership adopted the specifications by letter ballot, some forty thousand copies were printed and distributed.[40] As one ASTM member wrote in 1905, "There is hardly a work of any considerable size on which the physical testing of cement is not conducted," and the ASTM seemed to have cornered the market for guidance in this practice.[41] The recommendations were used by U.S. government engineers in the construction of the Panama Canal and on countless private building projects. They appear with regularity in handbooks and textbooks of the day.

Indisputably, the new regulations provided a solid basis for controlling the quality of concrete and cement. They outlined a series of tests that carried replicable scientific practice to a wide array of field sites, and afforded builders and building buyers the means of quantifying and regularizing a complex task. Simplified, portable tests fit perfectly with the idea of a "flow" process. Scrutiny of the ASTM specifications reveals, however, that the codes also called for practices that were neither quantifying nor regularizing in any direct sense. Within the lists of tests, sequences of specimen preparation, and delineations of desirable results we find both explicit and implicit demands for individualized judgment and good character on the part of the operator. Quality control, according to the scientific experts who wrote the specifications, actually required a blend of the uniform and the discretionary. With its new specification for 1917 (incorporating ASTM specifications), the American Concrete Institute issued this caveat: "In their use concrete and reinforced concrete involve the exercise of good judgment to a greater degree than do other building materials. Rules cannot produce or supersede judgment, on the contrary, judgment should control the interpretation and application of rules."[42] To follow the specifications as written—and thereby incorporate judgment into their quality-control operations—builders had to place testing in the hands of credible technicians: those deemed a priori to possess reliable judgment and appropriate personality.[43]

Creating a Conceptual Space for Expertise

The qualitative nature of the specifications—their call for technicians of high caliber—was reiterated on a number of levels. First, the ASTM's earliest twentieth-century regulations for cement and concrete testing devote much more space to describing tenets of testing than to listing specific test results. A desirable degree of fineness for dry cement is stipulated, as is the specific gravity indicative of pure samples, but in 1904 and in subsequent revisions of the text, few other absolute values are supplied. This stress on standardizing *methods* of test rather than results was in some sense a technical practicality. Research into the nature and behavior of cement and concrete was ongoing, and knowledge of acceptable strengths and the relationship of ingredients to strength changed frequently. While some criteria, such as the desirable degree of fineness for a cement or speed of set for concrete, changed slowly, some, such as tensile strength, could shift quickly. The editors of *Engineering Record* believed in 1896 that it was simply too soon to recommend numerical values because knowledge of cement and concrete was still limited; they believed that standardizing testing methods was a preliminary step towards introducing complete specifications.[44] The ASTM specifications of 1904 and subsequent revisions in 1909 and 1917 contained values for strength but cautioned that these could vary according to circumstance. Another reason for this concern for avoiding absolute values in specifications was that perfectly good cements and concretes could be produced from a variety of raw materials. But the focus on test method rather than on result also indicates that authors of standards and specifications envisioned the regulations as directives for the *work* of testing, not just for the finished product. Essentially, an emphasis on methods of testing created a conceptual space in which experts could define the behavior or qualifications of the ideal tester. They could thereby associate scientific quality control for cement and concrete with particular practitioners—specifically, their own students.[45] This association dovetailed with commercial pressures felt by building firms, as we will later see.

On one level, the experts filled this conceptual space with a general rhetoric of competence. Even as they recommended the reduction of testing to simple steps, experts claimed that testing required a "natural aptitude." This labor could not be reduced to steps that just anyone could perform.[46] The notion of natural aptitude was paired with its opposite, the idea of the "personal equation."[47] This was a fairly old conception that personal idiosyncrasy could disrupt a scientific procedure by introducing physiological or psychological variation on the operator's part, skewing observations or the operation of an instrument. By 1900, anxiety about the personal equation had moved outward to many technical areas, possibly spurred by the idealized notions of industrial work that formed the basis of scientific management. For materials experts and other proponents of the new standards and specifications who embraced the significance of aptitude or its absence, the efficacious use of cement was "dependent on the judgment and skill of the tester," not on the test alone.[48]

Had materials experts not celebrated the role of judgment in industrial testing, their elevated occupational standing might have reasonably been expected to drop. The idea that standards and specifications would be published and then widely used by engineers at work for industry suggested "economies of scale" incompatible with the world of intellectual specialization. Cutting and pasting published specifications into the documents of daily business does seem to intimate a reproduction of scientific practice, rather than its production, and it is not hard to see how the very dissemination of sample specifications could erode testing engineers' status in industry. Consulting engineer Richard Humphrey warned an audience of ASTM members in 1902, "It is the class of consumers . . . who draft their specifications with the aid of a pair of shears, that occasion the greatest trouble, not only to themselves but also to the manufacturers. Such persons as a rule with their limited knowledge of the properties of cement, select for their specifications the most rigid clauses from a number of other specifications, and the result is a new and impracticable standard."[49]

Many engineers therefore emphasized both the economizing features of standardized quality control and the necessity for intelligent use of these instruments. For example, Daniel Mead wrote in 1916 that there was "no better source of information than a well-prepared specification," and that rewriting a well-prepared published specification "in order simply to effect originality is not advocated."[50] Perhaps the very unoriginality of the published protocols seemed to reflect their grounding in an objective understanding of materials, safe from any personal deviance. Mead noted too that in writing specifications, "there are few needs for elaborate rhetorical effort . . . ; such efforts, even if they do not appear absurd, attract attention to the manner of expression rather than to the matter expressed, and are hence objectionable."[51] Stylistic flourishes are not merely inappropriate to the conveyance of technical information; they might actually disrupt technical work (here form *is* function). Yet Mead warned that freedom in the assembly of prepared specifications for a new project was vital: "Unintelligent copying of specifications is altogether too common, and is apt to bring embarrassment and shame on those who follow it."[52] Again, the qualifications and reliability of the tester were seen to secure the usefulness of the scientific procedures, however routinized they had become through their reproduction in published specifications. The ability to wield a scissors did not an expert make.

The emphasis on preexisting knowledge as a requisite for good testing fit nicely with engineers' more general conceptions of how science could best work for industry. In much of the prescriptive literature about engineering specifications during this period there is a sense of the mutability of scientific procedure, all for the good of effective testing. As they made clear in their instructional materials, many university-based materials experts saw formulas and tabulated data simply as an "anatomy" lacking the "flesh and blood of reality."[53] Real-life experience should shape the use of such technical tools. But at the same time we also find many engineers declaring that the abstraction of science is actually beneficial to testing practice. For Herbert Gilkey of Iowa State College, a "water-cement-ratio law" developed by Duff Abrams at the University of Illinois failed to address "important variations 'within the law.'" But this failure, for Gilkey, did not limit the law's fundamental usefulness as a general guide to practice, nor did it "detract from the credit that is due the man who through his researches first recognized the usefulness of the relationship and gave expression to it."[54] In fact, for a number of experts the necessarily "arbitrary" or artificial nature of tests and instruments was useful, because it did away with distracting features encountered in the actual application of a material. The translations enacted in the course of scientific investigations—for example, crafting cement into a standard form of briquette, or reducing strain to a calibrated measurement—make understanding the specimen a "doable" project. Abstraction is hailed for its potential to increase efficiency. In practical use, engineering knowledge is best contained in abstracted or digested form.

This formulation seems to join two incompatible qualities: scientific abstraction and industrial practicality. But it held no insurmountable contradictions for the materials experts or their industrial clientele, because they placed confidence in the tester's ability to translate general scientific principles into the specific terms of the job at hand. Regarding reports made of materials tests for industrial clients, a 1925 textbook recommends that "in general, it is best to eliminate all calculations from the body of the report. Methods and details of computation are more logically placed in an appendix to the report proper, where they may be consulted and checked when necessary. They are present as supplementary evidence without breaking the continuity of the report proper."[55] In such a scenario, a client can proceed from the findings of a test without careful study of its details. The workings of the tester's mind are to be trusted. Such recommendations helped carve out a particular identity for materials testing based on the personal characteristics of the tester.

Some of the arguments that experts offered for their distinction between test and tester seem no more than circular. Consulting engineer W. Purves Taylor, addressing the Association of American Portland Cement Manufacturers (later the Portland Cement Association) in 1904, claimed that the more important a given test was for ascertaining the quality of a cement, the more difficult it was to perform precisely and the more it relied on personal judgment.[56] He did not indicate, however, any way to measure the significance or difficulty of a test. Nor did he state whether the more significant tests were especially complex—and contemporary literature suggests that this was not the case—or offer any other technical reason for his claim. Similarly, Richard Humphrey, a member of the ASTM's 1904 committee on cement testing, invoked the skill of the tester as an almost occult quantity: "Since the experience of the operator plays such an important part, it would be impossible, therefore, to specify a single value for each requirement for tensile strength. The requirements of each engineer will naturally be based on his experience, and the committee cannot be expected to know just what those requirements should be."[57] Paul Kreuzpointner, an engineer of tests for the Pennsylvania Railroad (a major user of concrete), warned members of the ASTM in 1902 not even to seek a quantitative basis for testing: "In work requiring purely mechanical skill, the result can be measured by rule and compass, and these are tangible quantities. In testing, however, the results must be accepted in good faith, and a man's integrity becomes a leading factor of the reliability of work performed, because no one can measure or caliper the result."[58] But the circularity of these arguments does not diminish their historical interest. We can conclude from them that the experts' belief in, or at least their promotion of, the importance of experience and "natural aptitude" for effective testing was for many people in this field foundational, and that adaptations of testing for commercial

Ambiguity and Subjectivity in Specifications for Concrete

The ideology behind this general rhetoric is manifest in the content of the specifications themselves. From the ASCE's initial attempt in 1898 to craft standardized procedures for cement testing, specifications combined clear, inelastic requirements for certain steps in testing with much vaguer recommendations for others. As noted, in fineness tests conducted with sieves, the specification committee noted the amount of time a sieve should be shaken (1 minute) and the rate at which it should be shaken (200 strokes per minute). In contrast, the description of how materials should be mixed for briquette tests is notably ambiguous. Recall that this description has operators "vigorously knead" the sample with their hands "for an additional 1½ minutes, the process being similar to that used in kneading dough." While the directive notes that "a sand-glass provides a convenient guide for time of kneading," there is no way to associate "vigorous" kneading with any particular degree of strength.[59]

In its 1904 specifications based on the ASCE's work, the ASTM instructs the tester to form a sample of cement into a ball with the hands, "completing the operation by tossing it six times from one hand to the other, maintained six inches apart." Elsewhere in the specification the operator is told to draw a trowel over a filled mold with a "moderate pressure."[60] As specific as these instructions seem at first glance, enumerating and qualifying hand motions, the steps outlined would clearly not lead to identical physical treatment of specimens from test to test. These points of imprecision should not be construed as sloppiness on the part of the authors but as devices intended to consign authority for testing to people deemed to possess discretion and experience. The instructions were written by the same engineering professors, here working in their capacity as members of a professional society, who as educators also claimed that the property of discretion was a vital part of university engineering education. As indicated, paramount in the teaching of materials testing was the idea of limiting rote exercise and avoiding the excessive use of printed forms and calibrated instruments, in favor of coursework that inculcated experience.[61] The authors of specifications expected knowledge of what *vigorous* or *moderate* might mean to be part of young engineers' experience.

Given the popularity of ASTM specifications for concrete and cement—which remain the prevailing regulatory instrument of the industry today—and the persistence of their indefinite character, it would be easy to foreclose analysis of their social impact by simply noting that the specifications as written were "precise enough" to do the job of controlling quality. One might assume, for example, that hand motions of any strength or vigor would yield viable test specimens. But this sufficiency is itself historically contingent. How are we to know that vastly more precise instructions for cement testing would not have been required to achieve the same quality results if a different—that is, less trained—workforce had done the testing? Again, the content of the specifications, the particular blend of exactness and inexactness invoked, can be our guide to detecting their fuller social utility.

In 1911, for instance, the ASTM published specifications for the preferred tests for "constancy of volume" of cements. This type of test reflects the structural integrity of small cement specimens and is a good indicator of long-term performance. The authors recommended that three pats be made from the same batch of mixed cement and subjected to different conditions: one kept in air at normal temperature for twenty-eight days; one kept in water "as near 70 degrees F as practicable" for twenty-eight days; and a third "exposed in any convenient way in an atmosphere of steam, above boiling water, in a loosely closed vessel" for five days. To pass these tests of constancy of volume, the pats had to "remain firm and hard, and show no signs of distortion, checking, cracking, or disintegrating."[62] The variation possible under the terms of these instructions seems great. First, *firm* and *hard* are clearly subjective terms. Second, the ultimate reliance in this procedure was on the tester's eyes, resting on the notion that the pats had to "show no sign" of damage. Presumably, if those writing the instructions had not fully trusted the tester's eyes, a microscope or even a magnifying glass would have been called upon. Such a faith in subjectivity is evident throughout. Since these tests were to be conducted under field conditions, most likely in temporary sheds, and not in any kind of established laboratory setting, the concept of "normal" temperature is indisputably vague. "As near to 70 degrees F as practicable" is not much better.

And in fact problems in this area did arise. Ernest McCready, a member of the ASTM Committee on Uniform Tests, complained in 1907 that field laboratories with tin roofs, while "otherwise well suited for the purpose," were far too hot for viable cement testing.[63] On the other hand, he said, wintertime tests that left briquettes exposed to fresh air for twenty-eight days also yielded misleading results. Yet ASTM instructions for the volume tests did nothing to resolve this difficulty because the problem was seen to reside not in the specifications but in their use. As McCready himself explained, while uniform tests had been provided by the society, "adherence thereto . . . is still a long way off." He complained of the badly done tests that "the results are recorded without comment and at some time later are serenely compared and averaged with [those of other specimens] broken, perhaps in midsummer under exactly opposite conditions, and the cement gets all the credit or blame."[64] It is not cement that has wrought the test results, deserving credit or blame, McCready emphasized, but people, the testers. McCready is referring here to the kind of disputation that often occurred between providers of cement and their customers. The two groups commonly denounced one another's testing as flawed, and this frankly commercial aspect of scientific credibility will be explored below. But we might note that in finding fault with certain testers, McCready creates a conceptual space for other, "good" testers and overall weds the efficacy of science-based testing to the people undertaking the investigation.

Perhaps most strikingly subjective in the constancy-of-volume test specifications is the call for a steam testing apparatus to be set up "in any convenient way." A simple interpretation would be that a great variety of arrangements of steam and specimen would yield usable results. But in fact this was not the case. The 1911 specifications acknowledge the dangers of the steam, or "accelerated," test for constancy of volume: "The [accelerated test is] not infallible. So many conditions enter into the making and interpreting of it that it should be used with extreme care."[65] This dichotomy between articulated prescription and intended practice is telling. In the minds of the scientists, cement producers, and purchasers of cement writing these directions, adequate care could—and in fact was expected to—offset the vagaries written into the specifications. For good or ill, it is people, not machines or otherwise disembodied techniques, that bring science to bear on building construction. There was consensus that science and technology could not reasonably be expected to compensate for a lack of occupational pedigree.

The Limits of Mechanization

Controversy over the mechanization of sample preparation offers additional evidence of the testing engineers' special self-concept and of their approach to science as a somewhat "nontransferable" body of knowledge. The U.S. Army Corps of Engineers described the correct means of filling briquette molds in a 1901 overview of their testing procedures: "The tapping or ramming is to be done as follows: while holding the forearm and wrist at a constant level, raise the rammer with the thumb and forefinger about half an inch and then let it fall freely, repeating the operation until the layer is uniformly compacted by 30 taps."[66] The authors claimed that "the method permits comparable results to be obtained by different observers," but not everyone agreed. An engineer in the cement-testing laboratories of one major railroad complained that the corps' instructions "depended largely on how strong the operator felt at the time of making the briquette, which in consequence varied considerably in compactness and resultant strength." He invented a mechanical rammer that would regulate the pounding. Another railroad testing laboratory developed a mechanical mixer, which reduced "to the extent of mixing, at least, the personal equation in cement testing."[67]

However, the government's concrete engineers rejected mechanical rammers and mixers for testing work in this period, as did the ASTM in its cement-testing regulations of 1904. The government's engineers found that tests conducted with mechanically prepared specimens yielded artificially high results. The ASTM found mixing and molding machines to be unreliable and cumbersome. The operator of a mixing machine, for example, could tell if mixing was complete only by stopping the machine. This would hold up construction.[68] By 1909 the ASTM was still not obtaining satisfactory results with machinery. Was this because the society's experts did not have the technical knowledge to solve the problems of mixing or ramming machines? This seems unlikely, as sophisticated machines were then emerging for handling plaster, lumber, and even concrete. By 1908, mechanized "continuous mixers" for large-scale preparation of concrete featured automatic measuring devices for feeding sand, gravel, cement, and water.[69] Whatever mechanical actions were causing the artificially high results from machine-prepared cement specimens should have been identifiable and correctable. The problems of the mixing machines do not appear to have been insurmountable. It seems possible, instead, that the experts did not wish to mechanize their tasks completely but preferred to retain their unique capacity within a rationalizing production context. Simply put, testing could not, under this scheme, trickle downward to become a task performed by uncertified workers on the building site.

At this point we can tie together the occupational concerns of these technical experts and their sincere belief that other workers were unlikely to do a good job of testing. The qualitative features of testing, as defined by the specifications, connected testers to a set of personal characteristics routinely denied to other occupational groups on the construction site. A closer look at the organization of building firms in chapter 4 will elaborate this denial, but in the literature on specification writing the attribution of blame went hand in hand with social identifiers. The ethnicity of laborers was routinely included in anecdotes about their ineptitude. Much as Frederick Taylor used the heavily-accented "Schmidt" as an example of the tractable, thrifty "Dutchman" in his *Principles of Scientific Management,* an article by an engineering attorney on liability issues exemplifies an irresponsible cement handler as "Tony the Italian."[70] In contrast, the trained technicians—as we might designate those who performed tests, to distinguish them from the experts who prescribed test procedures—were certified by their own social and technical background; a certification begun in the university and continued here.

The availability of these practitioners to industry thus buoyed the status of materials testing as they carried it from the academic laboratory into the marketplace. With their own intellectual authority assured, they helped maintain the special standing of those experts whose regulations they followed. Simultaneously, the technicians performed a "rationalizing" operation on materials testing. They rendered cement and concrete testing affordable to materials suppliers and building firms, which it would not have been if only the elite experts themselves had offered the service. In short, technicians served as placeholders for scientific knowledge in the marketplace: visible and masterful, yet with an identity and authority constrained from above.[71]

"A Good and Workmanlike Job"

We have approached the standards and specifications that surrounded the use of concrete after 1900 with the assumption that the contents of these protocols, however vague and even absurd they may appear to modern eyes, had clear meaning to the people who wrote them. The absence of specificity—of quantification, of precision—was not accidental, nor was it necessarily likely to lead to inept practice. By their design, standards and specifications brought people believed to be competent to the task of quality control.

Even extremely vague protocols can be granted this utility. For example, the editors of *Engineering News,* in explaining how to avoid "indefinite specifications," made the following distinction. They cited as inadequate the following phrasing:

> Concrete shall be made of one part by volume of Portland cement, 2½ parts by volume of sand, and 5 parts by volume of broken stone.

The specification, they said, failed to acknowledge that "there is a great difference in the volume of cement depending upon whether it is packed in a barrel, shaken down in a measuring box, or merely cast loosely in to a box." Yet the editors' corrected version held this sentence:

> The sand and stone shall be measured when not packed more closely than by throwing it in the usual way into a barrel or box.[72]

The phrase "the usual way" immediately strikes the modern reader as hopelessly inexact. It is not hard, in fact, to find evidence that this very subject was a matter of extensive testing in the period: different sizes of stone, dropped from various heights, the resulting volume of material then measured.[73]

So how can the second version be considered an improvement on the first? This claim makes sense if we shift our outlook to that of the authors. They wished to delegate the task of using the specifications to trained personnel who could be presumed to be familiar with the idea of what is "usual" and acceptable in this and other procedures. Mead summed it up best when he criticized the phrase "the mixer must be of a size proportionate to the size of the batch mixed" as having "no apparent meaning . . . or at least no meaning which would not be covered by the phrase 'a good and workmanlike job.'"[74] In the minds of people creating and utilizing standards and specifications for concrete after 1900, that phrase, with its bluntly subjective wording, was as scientific, as useful, as any string of formulas or table of data—perhaps more so.

Conclusion

Accepting these written instruments as good-faith exercises on the part of their authors, and not attributing their ambiguities to accident or ineptitude, helps us understand their layered appeal for engineers and builders. We can start to see why the elitist aspirations of concrete-testing experts were not scuttled by engineers' and builders' concern—seen in other parts of the labor process—with achieving economies of scale. If classroom practices began to define the special competencies of industrial materials testers, then standards and specifications completed that definition and helped garner the assent of the commercial sector. The opportunities and rewards of a privileged employment status followed for concrete testers. Specifications buttressed this status by carrying a particular conception of concrete testing into a world of contracts and documents. Architects and engineers implied and encouraged the presence of trained testers' presence by evoking the new technical protocols. Standards and specifications, like the design of college coursework in materials testing, wed the control of materials to a delegation of credit and blame in the workplace.

These protocols highlight two important features of science as it unfolded in the industrial sphere, features that challenge assumptions about the objectivity of technical inquiry. First, it is clear that *precision* and *control* were not synonymous in the world of concrete testing after 1900. The actions that assured quality construction—measurement of a cement specimen's purity, determination of a concrete mixture's wetness, enforcement of performance standards—were often predicated on loosely defined criteria. This was not "bad" engineering by any means; the personal qualities of testing engineers were the basis of effective quality control. Because published regulations such as these are tools of communication, and presumably of uniformity, we may be surprised to find that they embed personal idiosyncrasy in the practices they seek to regulate. But it should be clear by now that anonymity in the assignment of industrial skills—the reduction of productive labor to small tasks that any person could perform—did not pervade every feature of physical work. Uniformity characterized the choice of testing personnel (consistently these were meant to be men of certain credentials), and it configured the end results of concrete testing (solid buildings). That many of the operations of testing that fell between those two points were *not* entirely uniform was neither inadvertent nor problematic.

We also have a sense now that the trust that engineers and builders placed in science-based testing was constructed from socially inflected ideas about technical skill, not from some notion that science does its work by eliminating all features of character or all worldly pressures. With such connections between technical practice and the day-to-day conditions of social and economic competition made evident, "scientific neutrality," as invoked by early-twentieth-century commentators, becomes a rather loaded proposition. The observation that Progressive Era business owners trusted technical experts is not the end point of any inquiry into the popularity of science, but the beginning.[75]

Chapter 3 continues this inquiry into the contingent meaning of scientific neutrality by probing a second "extratechnical" function for published standards and specifications: the demarcation of commercial relations and obligations among building trades. We have seen that the inherent subjectivity of ASTM or ASCE recommendations led engineering and building firms to employ college-trained testers. The utility of that subjectivity in the world of commercial exchange reinforced that commitment. The life of science-based quality control in commerce was a worldly one indeed.

CHAPTER THREE

Science and the "Fair Deal"
Standards, Specifications, and Commercial Ambition

In the supervision of work, the engineer becomes an arbiter. . . . He must see that justice is done.
—*Daniel Mead, 1916*

As technical protocols bolstered the supervisory status of engineers on the concrete construction site, securing a vertical structure of authority and opportunity there, they also helped control relations among commercial concerns—the lateral structures of competition endemic to the building industry. We may recall that many different occupations and businesses contributed to the erection of concrete buildings after 1900. Engineering and architectural firms designed structures; engineering firms supervised the selection and work of contractors; contractors performed the actual work of excavation, construction, plumbing, wiring, or painting; manufacturers and distributors supplied building materials. Standards and specifications, embedded in estimates, contracts, and civil building laws, regulated exchanges among these different specialties and among competing purveyors of each type of service.

The regulatory codes that defined the skills of trained engineers and managerial personnel in opposition to those of ordinary laborers on the site also controlled commercial opportunities. Their content paired descriptions of sound technical practice with delegations of responsibility and blame to particular practitioners on the construction site. The delegations, not surprisingly, generally favored whichever concern had issued the project specifications at hand: designer, engineer, contractor, or supplier. But most important, from a historical perspective, they did so by granting to scientific knowledge and practice—testing, inspection, and related techniques—a remarkable degree of influence in commercial affairs. As embodied in technical specifications, science might serve a business enterprise as both neutral adjudicator and powerful advocate. The standards and specifications surrounding concrete construction from 1900 to 1930 can show us the extraordinary complexity—and utility—of science crafted as commercial tool.

There were several ways in which standards and specifications carried distributions of credit and liability to the construction site. All are manifest in both the "model" specifications published by trade organizations such as the American Society for Testing Materials or the Portland Cement Association and in project specifications crafted from those models by individual engineering firms, contractors, or city agencies that commissioned concrete buildings. On one level, the regulations dictated quite directly which participant had the initial responsibility for a given task—such as testing cement specimens, laying a foundation, or erecting walls—and which was to be the final arbiter of the quality of that work. Specifications might name engineer, contractor, or supplier as the responsible party. Indications of who was to perform what work, who inspected whose work, and with what techniques, all found their way into these documents. Specifications might also employ a narrower category of delineation, indicating a required standard of performance for a material or a portion of the building. Most rigorous in the delineation of practice on the building site were project specifications that "named names"—calling for a particular brand of cement, prefabricated reinforcement, or waterproofing compound to be used on that structure.

All of these written instructions carried precisely defined legal and business obligations to a technical undertaking, and their power to confer or deny commercial advantage made them the subjects of constant dispute. In many project specifications we will see technical goals subordinated to fiscal agendas as engineering firms, contractors, and suppliers sought to control costs. Further, the rapid rise of city building laws after 1900 fueled disagreements between those who solicited and those who provided building services, with engineering and contracting firms often finding municipal requirements for safety and reliability in concrete buildings to be unfair and ill informed. Project specifications embodied in city codes and in the bids of providers became a site of such contestation.

On another level, the composition of standards and specifications for concrete addressed more general difficulties of using science to regulate commercial interchange. First, the building industry faced the seeming incompatibility of open-ended scientific inquiry, by common definition a disinterested intellectual enterprise, and commercial conduct in the free market, by definition self-interested. The mere presence of scientific procedures and quantified standards by no means reassured those who made and used concrete that they were obtaining a quality product. Buyers of a service or product fretted that any science-based claims made by its suppliers could be false. Sellers complained of biased conclusions drawn by buyers from buyers' tests. But the allure of scientific control was undeniable. In practical terms and as a means of self-promotion, the use of science-based testing could serve a business well. In specifications of the post-1900 decades we see an ongoing anxiety: How could science be exploited as a neutral determinant of quality, detached from biased application by material buyer or seller? For those who produced and handled concrete in this era, expertise bore no fundamental objectivity; the problem of dispelling mistrust was significant.

But the ostensible pursuit of scientific objectivity by industry raised its own challenge: the problem of *fixity*. Testing instruments and protocols, tabulated test findings, standards, and specifications all bring a fixed body of knowledge or practice to some feature of technical work. That fixity brought to concrete construction a reliability and speed it would otherwise not have attained, but it also brought distinct problems for the commercial firms seeking to profit in the building industry. To compete with other providers of a service or material, a business must adjust to market conditions. On a procedural level, specifications that decree a particular organization of work may make it difficult for a builder to use the cheapest labor or materials. On a reputational level, scientific test findings could associate a materials producer or builder with good results and bolster a firm's standing, but if undertaken without bias, tests left that provider vulnerable to negative findings, and thus to negative publicity. In short, the benefits of standardized quality control for concrete were great, but the risks were also large. We will see in this chapter many strategies by which commercial concerns controlled these risks through the composition of model and project specifications.

Businesses were abetted in these efforts to use standards for competitive advantage by academic testing specialists who worked both as co-authors of model specifications and as consultants to the engineering and building trades. In the latter role, testing and endorsing the products of contractors and supply firms, the academics mustered a notably protean definition of "good science." Matters as prosaic as payment and confidentiality were carefully explored; issues as broad as personality and competence—already raised in the educational setting—again come to the fore. These shifting interpretations of science all helped the construction industry regularize competitive pressures and embrace technical expertise with little fear of experts' judgments. This situation worked to the advantage of the academic experts as well.

Specifications and the Control of Commercial Opportunities

The voluminous written specifications that accompanied the erection of a concrete building after 1900 served in some very clear ways as channels of control among those paying for the project and those providing materials and services. If building specifications described the required depth of a concrete floor slab or diameter of a concrete column, building buyers and the engineering firms that represented them could reasonably expect a contractor to fulfill that requirement. If an engineering firm was using its own workforce to erect a building, its supervisory personnel could use specifications to direct the activities of the firm's foremen and laborers. The specifications served as a shared point of reference for buyers and sellers and were an important means by which a building owner or general contractor could assure that his investments were safely made.

But model specifications for cement and concrete, and for products related to their use, were most often issued not by associations of building buyers—say, factory owners—but by groups dominated by manufacturers of these building materials. The American Association of Portland Cement Users (later the Portland Cement Association) and the National Association of Concrete Users (later the American Concrete Institute) were instrumental in the creation of model specifications and the scientific data on which such instruments were based. As already indicated, the American Society for Testing Materials, the American Society of Civil Engineers, and many other large trade and professional organizations represented the interests of industry in the creation of model specifications. Such groups saw in science-based quality-control procedures, and other technical protocols, a means of managing the conditions of commercial exchange. Their efforts gave to science a powerful role in the marketplace, and a very particular form.

The efforts of the ASTM to define safe and efficient concrete construction were particularly ambitious and far-reaching in their impact. In 1903 the ASTM attempted to consolidate the many independent efforts then under way to create systematic specifications for cement. The organization brought together representatives of the ASCE, the Association of Railway Maintenance of Way Engineers, the American Institute of Architects, the American Association of Portland Cement Manufacturers, the U.S. Army Corps of Engineers, and the ASTM itself to form a "Joint Committee on Concrete and Reinforced Concrete." The ASTM's overt intentions were to systematize the huge quantities of test data that had been generated by these groups and to address "fundamental principles . . . to begin with elementary matters and to get at the subject from the foundation up."[1] Calling on eleven universities to assist with its investigations, the committee's initial forays covered such subjects as the behavior of beams, the placement of reinforcement, the shearing strength of plain concrete, methods of bending reinforcement bars, and the effects of loading and aging on concrete. The United States Geological Survey (USGS) also provided aid by making available to the project a "model testing laboratory" for cement built at the 1904 St. Louis Fair. In 1905 the work of the committee was divided among ten subcommittees, focused on subjects such as aggregates, proportions, and mixing; strength and elastic properties; simple reinforced-concrete beams; fire-resistive qualities; and historical matters.[2] By 1909 the joint committee had issued a substantial progress report.[3] In addition to design topics, the report addressed such logistical problems as the adaptability of concrete and reinforced concrete for different uses; the selection of materials; details of mixing and placing; and risks of corrosion by air, seawater, acids, and oils.[4]

By 1910 the joint committee had augmented its laboratory investigations with tests on actual buildings. The group issued its final report in 1917, after numerous interim reports and some internal conflict that resulted in the removal of findings about which ASTM members could not agree. Despite these omissions, the final report contained all of the recommendations mentioned above as well as formulas for the calculation of internal forces; allowable stresses for concrete in flexure, shear, or bearing (compressive strength); and factors of safety. It drew on data from French and German laboratories, the University of Illinois, the University of Pennsylvania, and other American laboratories. The specifications were finalized by the "Joint Conference on Uniform Methods of Tests and Standard Specifications for Cement," a three-member commission with representatives of the ASTM, the ASCE, and the U.S. government that had been formed to reconcile differences that had emerged among the participants over the years.[5]

Although the joint committee's final report—considered to have been officially authored by the ASTM—received some strong criticisms, it had much influence on the use of cement and concrete in America.[6] Many of its requirements remained unchanged from the joint committee's first progress reports in 1905 through the 1930s, when new high-early-strength cements with very different chemical compositions were introduced to the market.[7] As early as 1908, architects, engineers, and contractors used the ASTM specifications for cement and concrete more than they did specifications of their own devising.[8] By 1921, authors of widely used handbooks on building considered the ASTM specifications to be the "universally accepted" code for concrete. Preprinted standards and specifications for frequently used building materials were kept on file by building firms and inserted into contracts, and those of the ASTM were favored.[9] To understand the source and impact of this influence, it is necessary to look at the nature of the ASTM and other instigators of the standardization of American industrial production in these years.

Seeing itself as a clearinghouse, the ASTM established technical committees to review research on materials done elsewhere, and it published those results at annual meetings and in the form of transactions and specifications. In the first twenty-five years of its existence the organization issued 515 standards and tentative specifications, each initially published in an edition of six thousand. By 1928 more than fifty thousand reprints of separate standards were being sold by the ASTM each year.[10] The ASTM directed its efforts to monitoring and controlling the use of materials in industry rather than finding new uses for materials, because its essential agenda was to function as a go-between for those who produced and those who utilized materials, and between both such interests and the powers (private or governmental) that might legislate the conditions of material production and use (for safety or other reasons).

From its inception, the ASTM categorized members according to their place in the market: "producers of raw materials and semi-finished and finished products; consumers of materials; and a general interest group, comprising engineers, scientists, educators, testing experts, and research workers."[11] Each group, it was claimed, would benefit from the ASTM's work. Manufacturers would experience economies of scale and would be able to keep producing even during times of little demand, because the ultimate usefulness of their product would be assured. Consumers, on the other hand, could count on truly competitive bids from suppliers of standardized goods. They could routinize all aspects of purchasing and inspection and could look forward to greater uniformity and reliability in the goods and services they purchased. Importantly, standards would reduce the possibility of misunderstandings between manufacturers and consumers, a "vexatious and expensive" problem.[12] All in all, the arrangement seemed to offer a no-lose situation.

Over and over the ASTM stated its commitment to balancing the influence of its constituencies. Describing the composition of its technical committees, the ASTM claimed that "on committees dealing with subjects having a commercial bearing, either an equal numeric balance is maintained between the representatives of producing and nonproducing interests, or the latter are allowed to predominate with the acquiescence of producing interests."[13] The idea that the producer and consumer could be brought together on "equal footing" for their mutual benefit was a basic principle of the ASTM. The association made much of the fact that while it could not legally enforce its standards, the cooperative manner in which its standards were formulated would make their merit obvious to all comers. Yet for all its talk of a balance of power between producer and consumer, the ASTM operated under the auspices of its industrial members. Officially it ran on membership dues, but without the expertise, time, and facilities contributed by its industrial members, it is questionable whether it could have functioned to the extent it did. It seems likely that the specific scientific findings issued by the ASTM may have been shaped more by the programs of industry than by those of consumers, a condition historians have recognized in the standards put forth by other associations of the day, such as the American Society of Mechanical Engineers' boiler safety regulations.[14]

Before we cast the creation of industry standards as a device intended to benefit industry at the expense of consumers, however, two important points should be made to put the ASTM's position in perspective. First, methodical quality control was not a trend that, once put in motion, could be fully determined by any one party. The ASTM was encouraging a series of checks and balances that could adversely affect its own industrial members. Once alerted to the existence of standards, anyone could insist that they be enforced. For example, consumers who did not maintain advanced testing facilities could ask the National Bureau of Standards to corroborate or expand upon privately made tests. In addition, the bureau distributed samples of cements to the public at nominal cost, inspected the sieves with which the fineness of cement was measured, and evaluated the operators who used them. In the area of concrete construction the bureau also collated the test results issuing from its own laboratories and those of the USGS, publishing about five papers on cement each year. An informed consumer of cement could turn to the bureau to check the validity of an ASTM specification or the quality of a supplier's product.[15]

Second, the ASTM did not see a conflict of interest in its provision of scientific services for support of its industrial membership.[16] Like other professional organizations of the day, it associated its work with a dynamic integration of science and business, whereby intellectual inquiry, enacted in a spirit of general economic uplift, could only lead to universally helpful knowledge. In 1907 Robert Lesley wrote of the joint committee's recommendations, "The usual policy is to have manufacturers and consumers equally represented . . . but in this case it was eight manufacturers and twenty representatives of consumers, which I think speaks for the fairness and broadmindedness that have served to make this a real, living specification."[17] Lesley almost seems to be protesting too much. Why point out the preponderance of consumers on the committee unless the usual array of participants is unfair? But the crucial point here is that an ASTM specification was not a static entity but the result of open intellectual exchange (a "living" specification). There is little reason to suspect Lesley or other ASTM boosters of insincerity in their belief that science might level the playing field of capitalist exchange. The ASTM's promoters conceived of the sphere of producers and consumers as a community of interdependent participants, albeit one in which producers could achieve security and power.[18] The promise of science was as an engine of economic security and power—of advantage—fairly achieved.

Reconciling Identities: Science "versus" Commerce?

With its aura of fairness and the true economies it could bring to commercial practice, the promotional possibilities of science for the construction industry were clear. For trade organizations that performed scientific research on cement and concrete products, publicity about that work was as useful as the scientific findings themselves. Some of the most concerted efforts to wed scientific standards and practice with the commercial use of cement and concrete were those of the Portland Cement Association, another source of model cement specifications after 1900. The PCA originated in the informal meetings of Pennsylvania cement producers who had been having difficulty with the handling of the reusable sacks in which cement was transported. In 1902 the producers formed the Association of American Portland Cement Manufacturers to address "business matters of common interest, especially the problem of the return of cement sacks."[19] In the ensuing decade a double agenda of original scientific research into cement performance and wide-ranging promotional efforts took shape.

By 1916 the Association of American Portland Cement Manufacturers had established the first of its state-of-the-art laboratories, at the Lewis Institute of Technology in Chicago. That same year the organization renamed itself the Portland Cement Association, operating under the pithy slogan "Concrete for Permanence."[20] The PCA staff grew from 8 employees in 1916 to 40 in 1919 and 450 in 1925, of whom 300 were "experienced engineers." Its membership in 1925 consisted of ninety manufacturers of cement in the United States, Canada, Mexico, Cuba, and South America. Headquartered in Chicago, the PCA maintained twenty-nine district offices as well as separate departments for the study of cement use in highways, housing, farm structures, railroad structures, and large engineering projects such as industrial buildings, dams, bridges, and docks.[21]

In 1926 the PCA erected a new headquarters building in Chicago that included 19,000 square feet of laboratory space. In the first year of operation the laboratory submitted more than thirty-five thousand specimens of cement ingredients and mixes to some fifty thousand tests. The PCA's scientists investigated the effect of water/cement ratios on concrete strength; the effect on concrete of size and grading of aggregates; proper methods of curing; and the effect of different waterproofing compounds, impure water, and other deliberate and accidental additives. Their findings were published in technical papers and in trade and engineering journals and were distributed by PCA field engineers as they visited cement companies, construction firms, and other interested cement users. The information was incorporated into specifications for cement as well.[22]

Like concrete experts in other associations and agencies, officers of the PCA believed that the dissemination of model specifications for cement could encourage the successful use and popularity of concrete, but image was almost as important as actual improved practice. The PCA generated tremendous numbers of publications explicitly connecting cement with science. Scientific facts about cement found their way into newspapers and popular magazines as the PCA distributed its yearly *Editor's Reference Book*. Many editions featured pictures of scientists at work in the association's laboratory, the "Workshop of Science." That these materials were intended to enlist the lay reader is evident in such remarks as "You may never have thought of it, but the cement that goes into a sidewalk or farm feeding floor might have gone into the beams of a skyscraper or the trusses of a great bridge." Reassurance that "chemists keep constant watch" on all PCA-monitored cements also indicates that the association saw its work as capitalizing on and expanding the public identification of science with quality products.[23]

The conception of science as a neutral arbiter of quality and commercial exchange that was the basis of these advertisements did not go unassailed as testing procedures found their way into concrete construction. The challenges faced by the PCA are clear in one particularly dramatic promotional event. At the 1904 Louisiana Purchase Exposition in St. Louis, the PCA contributed to the creation of a fully operational Model Testing Laboratory for the mining and metallurgy section of the fair. The laboratory was designed by a committee of the PCA and operated by the American Society of Civil Engineers. Members of both groups, including cement manufacturers, editors of trade journals, and academically employed scientists, combined their expertise to create for the visitor a particular experience of concrete. The exhibit offered to the general public and visiting engineers an impressive array of new scientific techniques for assessing the quality and strength of concrete, and it presented concrete as a material of great technological sophistication and commercial potential.

Indoor displays held machines for mixing and molding concrete, state-of-the-art testing equipment, and a library of literature on concrete, cement, and mortar—all staffed by working cement and concrete experts. Visitors strolling outdoors could inspect an exotic structure: a short concrete stairway curving up to the end of a freestanding concrete beam. In surviving photographs, this cantilevered platform looks strangely disembodied or unfinished. It stood in stark contrast to the elaborate commemorative ornament common to the Beaux Arts–style buildings around it (fig. 3.1). Early in the fair the beam's 33-foot length, reinforced with embedded steel bars, had been subjected to heavy loads that simulated the conditions it would have encountered if erected as part of a building. We may today consider concrete a most familiar and mundane material, but this was not true for visitors to the 1904 fair. This was an immensely popular exhibit,

FIGURE 3.1. Cantilevered concrete beam erected beside the Model Testing Laboratory jointly operated by forty cement manufacturers at the Louisiana Purchase Exposition, St. Louis, 1904. The exhibit was extremely popular: the outdoor beam was subjected to repeated testing before the public. *From Richard Lewis Humphrey, "Results of Tests Made in the Collective Portland Cement Exhibit and Model Testing Laboratory of the Association of American Portland Cement Manufacturers," 1904*

and it won two grand prizes awarded by the fair's directors for "attractiveness and interest."[24]

The display was conspicuously labeled a "collective" exhibit, having been sponsored jointly by forty cement and concrete concerns and public agencies. Thus, it also presented concrete as a product that transcended pressures of the competitive marketplace. The laboratory tested a wide range of cement and concrete products, from cement mixes to reinforcing bars, and then published the results without using any brand names. Even the buildings themselves were built with a mixture of four different brands of cements, to dispel any suspicion of overt self-promotion. Concrete appeared here to be a material born of disinterested scientific investigation.

To the purveyors of cement and concrete products seeking a solidly scientific public image, an aura of investigative infallibility was as vital as their claims of investigative objectivity. In a discussion of the cement specifications under revision in 1907, Richard Humphrey warned his fellow joint committee members not to make trivial changes to the published specifications, because "if adopted," such trivial changes "would tend to throw doubt on the whole specification." One member who suggested amendments to the specifications was told to hold off, "for among other reasons the United States government has just accepted them, and we ought not to admit so soon that they need adjustment."[25] Most suggestively, a committee member proposed that "the changes that are suggested in the specifications should not be broadly termed changes. They are improvements. We, like others, are progressive."[26]

Certainly not everyone involved in writing the specifications felt so protective of them. But these sentiments reveal the importance their authors attached to the specifications as vehicles for enhancing their own credibility as technically informed and trustworthy practitioners. The very idea of progress in fact became a useful defense for the inevitable obsolescence that accompanied much of the technical study of concrete. Herbert Gilkey, recording some years later the labors of his mentor, Anson Marston, nicely indicated that even outdated technical achievements merited respect. He refers here to Marston's development of modern concrete highways for Iowa: "The brick pavement of Marston's earlier researches was soon to become a relic, a link . . . in the chain of progress towards today's superhighway."[27] Marston's technical contributions are clear, but we can nonetheless appreciate the sense of intellectual sturdiness such a description lends to knowledge subjected to rapidly changing demands.

Academics as Industrial Consultants

The expansive language of the PCA's brochures, the theatricality of the St. Louis exhibit, and the joint committee's concern with a progressive public image are expressions of commercial efforts to sustain a scientific profile for concrete. But the conceptualization of concrete as a product of unbiased scientific investigation shaped the most esoteric technical work as well: the testing and research services provided by academic engineering faculties for the building industry. Here, tensions between the supposed neutrality of scientific investigation and the self-serving operations of industry saw further manifestations. As was the case for their industrial clients, it was in the interest of academics to avoid any hint of impropriety. But again, the financial and reputational stakes of materials testing were high, and the work of testing did not lend itself readily to an air of intellectual detachment. Because we have already seen their self-conscious reputational efforts as instructors of a technical discipline, it will not be difficult to trace the means by which academics achieved a particular occupational identity as consultants to industry.

We have already seen that many universities made commitments after 1900 to serving industry. The establishment of engineering experiment stations in state universities and extensive testing laboratories in private schools grew from the idea that education and service could form a desirable pairing in an era of industrial expansion. We can now start to see that that compatibility had a rather layered nature. In part, the idea of providing services to industry came from a desire to promote the general field of engineering as an area of intellectual specialization. Anson Marston of Iowa State College, delivering a talk on "the engineer's responsibility to the state" in 1931, pointed out in his characteristically grand prose that the engineering profession "differs from other learned professions in that it is essential to the very existence of our modern industrial civilization.... Today's great problems of civilization center around the relations of the producer and consumer."[28] It seems likely that the actual subject of this talk may have been "the state's responsibility to the engineer." At many land-grant schools, engineering instructors pointed to the successful model of service to agricultural interests, thinly disguising their resentment at the secondary treatment of engineering by government funding sources. As Arthur Talbot of the University of Illinois wrote in 1904,

> While some people have been looking down upon the farmer as a man who kept his nose to the soil, I am of the opinion that the farmer, with the Department of Agriculture and sixty experiment stations employed and paid to solve his problems and tell him what to do, has a decided advantage over the engineer who is performing haphazard experiments in the making of concrete, and over the mechanic who is forced by his ignorance to keep his nose to the grindstone.[29]

Talbot's predecessor at Illinois, Lester Paige Breckenridge, had also worked long and hard to carry the university's existing service role to agriculture over to the world of manufacturing and construction. He believed that "systematic research and service" would be complementary to the primary mission of the College of Engineering: "the education and training of the youth of the state to become honorable, useful, and successful citizens in the conduct of public and private enterprise." Historian Charles Rosenberg has suggested that the agricultural stations received official sanction in the 1870s because they would help rationalize and systematize farming operations, "providing, that is, a conservative alternative to more radical schemes for adjusting to changed economic and demographic realities."[30] Thirty years later the promoters of engineering experiment stations still used socially conservative rhetoric, touting the creation of useful citizens and the support of free enterprise. They simply attached their offerings to current economic conditions, serving manufacturing industries rather than farming.

On this foundation of social and economic importance, university faculty who specialized in materials testing embarked on a program of service to industry carefully designed to lend industry the authority of science without undermining industrial prerogatives. Working within well-funded laboratories that they had helped design, the University of Pennsylvania's engineering faculty offered testing services to industry and set up a network of business connections based on advising and endorsement. All testing was conducted with an eye to the department's business utility. When they were unable to provide the testing facilities desired by inquiring clientele—who ranged from small local manufacturers to Bell Telephone to government agencies—Penn's engineering faculty behaved like good businessmen and sought alternative means of serving their clients. In 1911 a manufacturer of lime for cement wrote to H. C. Berry asking if Penn's engineers could determine the effect cold would have on setting cement made with its product. Berry replied that while Penn itself did not maintain a cold-test facility, he was confident that a nearby cold-storage company would accommodate such a test conducted under his direction.[31] A similar business acumen figured in the laboratory's relationship with local commercial competition. Penn's engineering instructors would turn down jobs they thought to be too routine on the basis that this refusal left them available for testing projects of a more unusual character. It also kept relations friendly with local commercial testing companies.[32] Penn's engineers and Philadelphia companies frequently passed clients back and forth. Each laboratory could count on the other to take care of clients it could not accommodate itself and to pass along clients better suited to the other's capacities.[33]

Relations between the academic engineers and private providers were not always without conflict. Anson Marston was the subject of a lawsuit in 1909 in which he was accused of infringing on patents held by the Cameron Septic Tank Company for concrete septic systems. The plaintiffs felt that academics often held themselves to be "above the law." Marston had designed septic systems for several local municipalities and Iowa State College's Ames campus, and such disputes no doubt worried his employers. Land-grant schools were dependent on the goodwill of state legislators for their operating budgets, and those legislators in turn sought the approval of private enterprise. In the years following this suit the college inaugurated increasingly rigorous methods of patent review for its faculty.[34] But for the most part the academics moved easily between proprietary and cooperative work. In 1910 H. C. Berry of the University of Pennsylvania conscientiously marketed the portable extensometer, which measured cracks and strains in concrete, that he had developed for his students' use. He wrote to several instrument companies, seeking the greatest profit on the extensometer's production, but in the same year gave free copies of plans for the instrument to engineering departments of several universities and the federal government's Watertown Arsenal.[35] Berry and his colleagues also gave advice on laboratory design and interpretative data to other universities without hesitation or charge.[36]

What was the basis for this kind of decision—when to share freely, when to withhold or charge for knowledge? It may be that market value determined the actions of the academic consultant. Berry's extensometer could have brought him acclaim *and* profits, while ideas about laboratory design might not have been a potential source of revenue because universities did not customarily pay for this kind of advice. But a subtler set of criteria for proprietary decisions was also at work, as a close look at laboratory acquisition and endorsement practices at Penn and other service-oriented departments reveals.

The laboratories of the University of Pennsylvania engineering school acquired materials for testing in several ways. Records show that at times the university would order and pay for samples of commercially produced building materials. In other instances companies that made cement, reinforcing bars, or concrete sealants might send free samples to Penn, asking for test results in return. On other occasions, depending on their curriculum, faculty members might solicit free samples from companies, offering in exchange to submit test results to the supplier. These interactions were surrounded by rules of conduct. Penn's faculty members explicitly said that they "did not wish to become involved in advertising" and that test results must not be used in this way.[37] Yet in certain circumstances the favorable reputation of Penn's laboratory was readily lent. At the request of a former student, H. C. Berry agreed not only to test a specimen from the Honest Reinforced-Concrete Culvert Company but to include in the paid service a photograph of the Towne Laboratory's 600,000-pound Olsen testing machine at work, clearly identifying the specimen under test as that of the Honest company.[38] In other instances Penn would publish press releases that served a similarly suggestive function.[39] In essence, Penn's reputation could be attached to a particular commercial product, even if actual test results were withheld. Penn faculty members shrewdly treated their endorsements of a supplier's product and test results on that product as *two different commodities*. This demarcation was ingenious and was crucial to the academics. It seems to have saved them from charges of conflict of interest, defining the university as a source of disinterested expertise while creating a sizable "market niche" for scientific knowledge.[40]

For related reasons, the issue of providing services without charge was complex. The University of Pennsylvania laboratory sometimes charged for its services—in 1910 a large railway paid the university $1 per specimen tested; Bell Laboratories was charged at a daily rate of $25—and sometimes did not.[41] Bartering testing services or implied endorsement for materials was common, and academic engineering departments were glad to use such trading to save themselves the cost of buying classroom supplies.[42] The Flexol Company, a maker of waterproofing compounds for concrete, received Berry's results of tests on free samples it had sent. Yet the City of Detroit was rebuffed in no uncertain terms when it requested free advice from engineering professor Edwin Marburg on wording for its concrete building code.[43] We might suspect that Marburg's reluctance to work without pay had to do with economic practicalities or with avoiding the liability attached to casually advising on a building code, but it also reflected a real ambivalence within the profession. On the one hand, the involvement of money in testing seemed to carry a taint of poor character. It is of course difficult to trace the incidence of bribery and other illicit exchanges in the world of industrial testing, and loud protestations such as that of Charles Dudley against "the hydra-headed monster, graft" might have been as much a gesture of self-promotion as one of genuine concern.[44] But many schools took the issue seriously. Anson Marston went to some lengths to explain that Iowa's engineering experiment station always charged clients only the "bare costs of the work, and minimal expenses."[45] In 1912 the president of the college also felt strongly that professors should not charge for lectures given to farmers and industrial organizations in the state.[46]

On the other hand, evidence suggests that to work without payment, especially on a large-scale project, could undermine an engineer's credibility. The editors of *Scientific American* wrote in 1920 that educational institutions researching industrial problems must charge for such services or risk their reputations: "Purely philanthropic enterprises do not engender in the managers of industry that confidence which is necessary for success. There must exist in the scheme of cooperation between the educational and the commercial establishments an element of mutual obligation and mutual benefit."[47] Apparently the failure to charge for services could imply amateurism or impropriety even if the occasional trading of favors did not. This concern played out differently in government research agencies concerned with materials research (such as the Bureau of Public Roads), which saw free service as part of their mandate.[48] However, within the private sector, payment for service was a calculated element of the materials experts' scientific reputation.

Trade Journals as Knowledge Brokers

As these examples suggest, the dissemination of scientific knowledge about concrete, like the production of that knowledge, occurred in circumstances shaped by competing interests. Any business related to concrete construction could marshal scientific findings for its economic or reputational benefit. As links between scientific investigation and commercial application, journals played a tremendous part in constructing the scientific reputation of concrete. Their editors and publishers operated with no less self-consciousness than the experts whose work they published, balancing an aura of objectivity with commercial intent.

The growth in the number of publications devoted to the production and use of concrete after 1900 was dramatic. In addition to transactions and proceedings published by professional and trade societies, including the ASCE, the ASTM, and the American Concrete Institute, a large number of privately published periodicals appeared with the boom in concrete construction. Some were devoted to the testing of materials, some to cement itself. *Cement Age* began publication in 1904 as a "clearinghouse for information on cement."[49] It offered summaries of cement research, descriptions of new products, book reviews, career notes, and, in its "Briquettes" column, notes on new cement companies. By 1911 it had merged with *Concrete Engineering*. The expanded publication announced its commitment to the many factions involved with cement production and use. For cement manufacturers, it would cover new techniques of cement use. For architects, it would address the problem of "artistic expression" with concrete. Finally, for engineers and construction specialists, the "men who are doing the 'big work' and are active in this 'great uplift' for *better construction,*" *Cement Age* would offer "detailed, specialized articles on methods and practice."[50] *Engineering Record, Engineering News,* and more specialized journals for municipal or chemical engineers also covered the field.[51]

The journals were a forum for the exchange of technical information, but because they were also commercial entities, they were rarely passive vehicles for the publication of scientists' work. In some instances journals directed researchers' work: *Concrete Engineering* requested that one author from the University of Pennsylvania provide additional test data for a particular article and called for an entire series of tests from the Case School of Applied Science.[52] In other instances editors seemed to direct the application of scientific work towards particular commercial ends. Robert Lesley, in his capacity as publisher and editor of *Cement Age,* complained at length about the "torture" to which some investigators subjected cements with accelerated tests (in which cement specimens were submitted to extremes of temperature and moisture). Lesley was also a leading manufacturer of cement, and if his publication advocated the abandonment of rigorous testing, it may have been to the benefit of his production processes and profits.[53]

There were certainly points at which the professional agendas of the journals' publishers and scientists overlapped. *Cement Age* editors advocated the awarding of state funds to university materials science laboratories. Allen Brett, the editor of *Concrete Engineering,* established a "Consultation Department" column in 1910 that offered industrial problems for students to solve while, he promised, "putting professors' names before thousands of men."[54] But the academics did not necessarily accept the journals' definition of editorial neutrality. Edwin Marburg, and no doubt many other instructors, relied on the book reviews in *Engineering News* and *Engineering Record* to facilitate the selection of textbooks for Penn's engineering courses. At the same time, Harwood Frost, a prominent Chicago construction engineer, fretted about the "abuse" of book reviews. Many journals were published or edited by materials producers, and Frost feared that a journal's editors might praise books from whose popularity they would derive profit. He planned to write a booklet on "reading between the lines" of published book reviews.[55]

The journals themselves sought on occasion to deflect criticism about their commercial involvements. In 1900, *Engineering News* distinguished its policies from those of journals that published favorable editorials about materials producers and then solicited orders for reprints. Over the following decade the periodical's editors regularly pointed to the "unscrupulous" practice of journals for building materials other than concrete that were wont to give "misleading" and "lurid" coverage to failures of concrete structures. *Cement Age* shared the view that technical journals should not be "organs" for particular products: "A journal best serves its readers and best serves the industry it represents when it presents the truth and when it treats all subjects from a broad and impartial point of view."[56]

This claim may strike us as disingenuous, coming from a journal that clearly had commercial interests. However, even if we do not accept this statement as literally representing intentions of honesty, it is still a reflection of how a professional sector wished to position itself. The journals were commercial entities, but their claims of scientific impartiality aligned them with a realm of pure inquiry. Such claims also fit the journals into the normative moral codes of the academics who provided them with materials for publication. Like the engineering instructor-consultants, the journals did not pretend to be entirely scientific. They readily paired test data with endorsements of brand-name products, because while they wished to promote the idea that academic inquiry was disinterested, they also wished to convey that it had important commercial applications. This was the same duality on which the academics constructed their own occupational security. The professional conduct of the commercial publishers and that of the scientific sector were mutually reinforcing. Such conceptual allegiances have much to tell us about the dissemination of scientific knowledge and practices in a commercial environment.

Networks of Favor: The Problem of Brands

The line between the exercise of technical expertise and commercial promotion was a blurred one. The academics recognized the promise and danger this blurring carried for their own discipline. Journal editors saw and manipulated the line for their own purposes. A calculus of liability and patronage could configure any application of science. As specifications took form through the work of various trade associations, a debate emerged about whether or not their users should amend them to include brand names. Should agreements between builders and owners specify the exact product to be used by contractors or, instead, indicate the nature and level of performance of materials, leaving the choice of supplier to the contractor? That is, would a contractor be allowed to select a favored brand of cement or reinforcement, or would that choice be determined by the building's engineer? A contractor's or engineer's favored company might well be of the highest quality, but who was to decide which companies would have the commercial advantage on a given project?

The blending of technical and promotional information in construction operations was well under way by this point. It was common practice after 1900 for suppliers to issue "facsimile specifications." These were essentially boilerplate texts, usually inserted at the back of a catalog or promotional brochure created by the supplier, and were intended for direct reproduction in the estimates and contract prepared by building designers or engineers. They carried descriptive information such as dimensions, design, strength, or type of material—be it cement, reinforcing bars, a prefabricated chimney, or a refrigeration system under consideration—but they also automatically inserted the supplier's brand name into a project specification when designers or engineers cut-and-pasted them into place.[57] In bundling technical information and brand names in this way a supplier assured the use of its products. If engineering firms in preparing specifications for a concrete building included a brand name, they were similarly guaranteeing the use of a particular product. Contractors, linked to their own, usually very local, networks of suppliers, objected strongly to this practice.

Part of this battle centered on the implicit contradiction of using standardized products and specifications in a supposedly open market. The ASTM frequently fielded complaints from materials producers who feared that they would be unable to achieve economical production if constrained by standards.[58] Robert Lesley and Henry Spackman, who owned cement and cement-testing firms, respectively, argued in 1908 that if specifications for cement were going to proliferate, then it was only fair that there should be as many rules governing the grading and quality of sand, another ingredient of concrete. They went so far as to draw up a chart showing the unequal number of words in contemporary specifications for cement and sand.⁵⁹

Manufacturers also worried that standards would interfere with competition. For example, in 1911 the Department of Commerce and Labor organized a conference of government engineers to review and consolidate specifications for cement used by all government departments. The engineers concluded that every bid for furnishing cement or doing work in cement must state the brand of cement proposed to be furnished and the mill at which it would be made.⁶⁰ Architectural journals of the day echoed this policy, but cement producers vigorously objected. The editors of *Cement Age* argued that in any specification that included brand names, the specification's writer arrogates "to himself the function of engineer, architect and purchasing agent" and thus severely handicaps the builder. They fretted that specifying brand names would lead to higher building costs and other unfortunate consequences for the building industry: "By unnecessary arbitrary decisions, before the work is started, doors are slammed and bolted in the face of all competition."⁶¹ Consulting engineer Daniel Mead pointed out in 1916 that the call for particular brand names could save money by eliminating the need for tests of a cement or related product.⁶² But manufacturers saw little benefit in that economy. They preferred a scenario in which any product had the fullest number of possibilities for continually remaking its own reputation and increasing its market share.

The use of very precise specifications could therefore carry broad economic effects that were not to everyone's liking. This perceived inequity could be cast as simply a by-product of capitalist competition: for an engineer to call for a brand of cement is conceivably merely an expression of a consumer preference. But the inclusion of brand names in specifications seemed to many manufacturers beyond the bounds of fair trade. Even aside from the issue of brand names, the authorship and content of specifications held implications for the distribution of responsibilities and, concomitantly, status, on the construction site. The editors of *Cement Age* continued their complaint about the "dangers" of specifications, here focusing on their adoption by architects: "The endeavor to further centralize all authority in the hands of the architect will relieve the constructor of more responsibility and will tend to make him simply an assembler of materials."⁶³ The inclusion of brand names reduced even further the contractor's autonomy. But this degradation of builders' workplace authority was precisely what some of the framers of standards and specifications had in mind. As the subjective nature of specifications helped elevate engineering skills above those of other laborers on the construction site, so the specificity of these directives could help grant engineers authority over different trades at work on a building site.

Some arguments about engineers' superiority were no doubt predicated on the idea, described earlier, that engineers were simply men of better character than members of other occupations. But advocates of technical professionalism also claimed an opposite feature of engineering practice—engineers' adherence to an established body of practice (a veritable suppression of personality)—as a reason for industry to rely on them. Prescriptive literature by experts in construction management heralded the notion that the field engineer's "first duty" was to the specifications, allowing neither deviation from the prescribed process nor substitution of materials. As one handbook put it, "Once they are agreed to by owner and contractor, the engineer loses control over plans and specifications, and instead . . . he is himself controlled by them. An obvious exception is the case where the contract expressly reserves the right for the engineer to make minor changes or to accept substitutes; but such an arrangement is fraught with dangers."[64] As in the case of pedagogical arguments, the apparent contradiction between praise for engineers' discretionary talents and for their obedience to established regulations posed little problem here. Specifications were meant to embed the authority of their creators, stratifying the skills involved in construction. Their effect on the workplace was to institutionalize authority, much as other bureaucratizing techniques systematized work relations in factories and offices.[65]

Building Codes

While cement experts, materials producers, engineers, and contractors shaped written instruments that would regularize their own relations, they were also faced with a sudden increase in the number of laws and statutes issued by city building departments. These building codes represented the public interest in a newly systematic way by legally calling for safe building practices, and in doing so they introduced a powerful set of institutionalized constraints into the competitive struggle for technical authority already unfolding in concrete construction. But city codes were not an entirely new or exogenous factor in the commercial development of concrete.

They were often based on the specifications of the ASTM, the PCA, or other organizations, or were compiled with the advice of those bodies. Nonetheless, they introduced into concrete construction a series of uncertainties and negotiations with which concrete experts and business people had to contend. The creation of city building codes became another arena in which members of the construction industry vied for authority and economic power, and in which the precise applications of concrete were formulated.

The legal codes governing building in turn-of-the-century American cities were of two kinds: fire codes and structural codes. Fire codes had existed since ancient Rome. Twelfth-century London, with its densely packed and highly flammable buildings, instituted laws requiring that space be left around fireplaces, that houses keep ladders ready for fighting fires, and that in the summer months barrels of water be kept handy. New York began regulating construction methods and materials to prevent fires in 1625. The expansion of insurance industries followed closely on these developments, with the first such system, complete with its own private firefighting brigade, established in London in 1680. During the nineteenth century, insurance companies proliferated in Europe and America. In the 1860s the United States saw the creation of the National Board of Fire Underwriters, intended to control rates and competition among American companies.[66]

Efforts at fireproof construction had been undertaken throughout the nineteenth century, many of the attempts oriented towards protecting wood or steel members with such noncombustible materials as plaster, hollow terra-cotta tiles, and concrete. Changes to the overall form of buildings also emerged, particularly for commercial structures. Elaborate towers or mansard roofs in which fire could expand disappeared in favor of simpler contours, and slow-burning mill construction—based on the use of heavy timbers—proliferated. Study of the behavior of buildings during fires also led to new understanding of how heating and illumination systems could be most safely designed.[67]

Fire tests that recreated the conditions of burning buildings were considered to be the most reliable way to learn about the combustibility of materials. In 1893 the National Board of Fire Underwriters established its own laboratory at the Armour Institute of Chicago.[68] With elaborate installations, the laboratory staff determined the fire-resisting properties of different types of construction and issued a series of recommendations based on studies of how the contents of a building affected the way it burned. The board's findings on fire load, as this concept was called, were brought out in 1905 as the first model fire code and were rapidly adopted by many city building authorities. Some thirty thousand copies were distributed between 1905 and 1922, when a revised edition was presented.[69]

By 1900 the fire-resistant qualities of concrete were being heralded. A 1902 fire at a New Jersey borax plant, in which steel-framed buildings collapsed while reinforced-concrete buildings survived, was widely publicized by concrete promoters.[70] By 1922 the National Board of Fire Underwriters' building code had approved reinforced-concrete construction for all types of building. This code was based in part on the ASTM's specifications for reinforced concrete, and it stipulated materials and testing procedures.

Ideas and attendant regulations about structural features of buildings emerged more slowly than did those regarding fire. It was not until the elastic theory of structural design was accepted in the mid-nineteenth century that methods of structural design could be systematized. The British Board of Trade was the first to undertake this task when it specified a maximum permissible working stress for structural members.[71] The board's determinations were derived from findings on the ultimate strength of wrought iron, divided by four to provide a margin of safety. The determination of margins, or factors, of safety—sometimes derided as "factors of ignorance"—was a complex task. As one 1906 editor put it, buildings once occupied are subjected to "distributed loads and concentrated loads, quiescent loads and moving loads, shock, wear, corrosion, freezing, and other forces of attack."[72] Given this daunting complexity, as architectural historian Henry Cowan has pointed out, materials that were studied in laboratory situations were often assigned a much higher factor of safety than those assessed in actual building site applications. Working loads—those loads a building would experience when actually occupied—were also difficult to estimate, because it was unclear how much of a building would be occupied at any time.[73] But by 1900 most city authorities had proposed some factor of safety for each major building material. For reinforced concrete, the factor was generally between 2.5 and 5, depending on the degree of compromise reached between those seeking design safety and those seeking economy.[74]

Historians have described almost constant change since the late nineteenth century in the way the writers of building ordinances have interpreted allowable loads and the characteristics of materials. New knowledge emerged from testing programs, and greater leniency in regulations occurred with the increased confidence that came with refined methods of determining design strengths.[75] New procedures and ideas moved outward from individual test sites into city regulations and from the codes of larger cities such as New York and Chicago to those of smaller cities. In 1900, New York and Chicago lowered standards for buildings on the basis of reduced interpretations of live loads, and like later changes, these new ideas found their way into broader practice.[76] A drive for uniformity in building ordinances also emerged after the turn of the century. The International Society of State and Municipal Building Commissions was organized in 1903 to address this issue.[77]

Building Codes as a Competitive Tool

As was the case with manufacturers facing the standards and specifications issued by the ASTM, building trades wanted to guard their competitive prerogative against what they considered to be excessive regulation under municipal codes. Each group had its own particular anxiety. Cement producers were wary of governmental efforts to create building regulations. Some saw the codes as making "extravagant demands" on reinforced concrete: "Building inspectors and many engineers have demanded from it a factor of safety twice as great as that required from steel, thereby making the cost prohibitive. Cases have actually occurred where an inspector has required that floor areas be loaded with six times the working load. . . . If the floor stood the stress the building was accepted; if it collapsed, the whole structure was condemned."[78] Others complained that "the average municipality calls in people without experience to make code, and then good engineers are forced to compete with others, who could follow those specifications literally and thus build more cheaply."[79]

Contractors also fretted about maintaining their control over the day-to-day construction process. If codes caused them to lose the right to determine which materials and methods were to be used, they could be trapped into supplying services at a cost they could neither predict nor control. As one contractor wrote, "Competency is secured by study and experience. . . . It is not necessary to go into such details as cautioning against freezing mortar. There would be no end to these detailed requirements if once begun."[80] Contractors referred approvingly to the "designer's initiative," taking the side of architects and engineers—whose prerogative they might at other moments oppose—to make the case against excessive government regulation. Builders' strategies for accommodating legal controls could reach a fairly subtle level. Some representatives of the building trades proposed the idea of introducing greater criminal responsibility into building, in lieu of stricter codes. In this way they would maintain their control of the workplace but would encourage careful work through the use of a negative incentive.[81]

To understand the negotiated nature of these codes, we should also look at the contestation that occurred within city offices as those bodies faced the demands of their different constituencies. Two cases briefly illustrate the power of codified advanced technical knowledge in the potentially lucrative commercial area. In Chicago the chief tester of cement lost his job in 1908 when he refused to accommodate certain suppliers and thereby fell out of favor with the mayor, with whom those suppliers had close relations. The city had maintained a testing division in its Department of Public Works since 1897. It tested all cement, concrete, brick, coal, and oil requisitioned by the city. By 1908 the division had six employees, including two metallurgical chemists and one specialized cement tester. Peter McArdle, the chief tester and a cement specialist himself, was demoted by the commissioner of public works after complaints from local companies that he had implemented unfairly rigorous material specifications. After a protracted court case, McArdle was reinstated with back pay. A local newspaper summarized McArdle's case: "He is an honest man. . . . As such, he was an indigestible morsel for the [Mayor] Busse grafters."[82]

In New York City at about the same time, "concrete interests" complained at public hearings that local codes granted too much power to city superintendents of building—there was at the time one such figure per borough—by allowing these officials discretion that might work against a "fair deal for concrete." Working on the New York City code at this time were not only representatives of the New York and the National Board of Fire Underwriters but also members of the American Institute of Consulting Engineers and the American Institute of Architects (AIA). These powerful and established organizations were somewhat above the commercial fray, so it is perhaps not surprising that representatives of the AIA believed that granting such discretion was fine. They believed that the code had been devised with the idea that there "would be honest superintendents. If it proves otherwise, we shall have ways of getting [them] out." In the end, however, the mayor put through a version of the code that did indeed restrict superintendents' authority. As historian Henry Comer has summarized, the formulation of New York's codes was an ongoing battle of "legislation versus administration, law versus discretion, and democracy versus autonomy."[83]

The Problem of Scientific Fixity

As is by now obvious, in enlisting scientific techniques and standardized protocols for concrete construction, the building industry had unleashed a powerful commercial tool, capable of helping or hindering those who trafficked in the new technologies. One of the most insidious threats that the new practices posed for building firms was the possibility that the sheer authority of science—its reputation for objective investigation—could simply turn against an enterprise, bringing to it negative testimony rather than endorsement. Science-based testing, often demanded by project specifications written by engineering firms, thrust a brand of cement or a construction firm into the harsh light of scientific investigation. A testing procedure, instrument, or performance standard subjected materials to definitive assessment, and for all the networking and mutual support occurring between technical experts and businessmen in the field of concrete, the facticity of science could still hold fearful consequences.

Manufacturers of cement and related products were most often subject to this kind of competitive "danger" as they worked to meet the specified requirements of engineering or contracting firms on a project. As the field of competing suppliers grew after 1900, cement makers also needed the flexibility to adjust their product lines and selling points according to market conditions. Definitive scientific characterizations of a brand could constrain that flexibility. Resisting scientific tests was not an option. No architect, engineer, or building buyer would knowingly choose a supplier who rejected quality-control guarantees. And scientific testing did serve important technical and legal functions—for all participants in the building industry; one of the advantages ascribed to uniformity in methods of cement testing was that it would "do away with a constant source of friction between manufacturers and consumers in regard to the results of tests."[84] Instead, manufacturers mustered a variety of constructive responses to the problem of scientific fixity—protecting the authority of technical expertise while evading its punitive potential.

First, a commercial firm could simply try to keep test results from publication, citing the unethical nature of such disclosures. William Steele, one of the largest builders of industrial concrete buildings in the East, agreed in 1913 to supply Penn's laboratory with a concrete cylinder for experimental testing only if results were guaranteed to remain confidential. Failed tests of cement or concrete could be enlisted by rival companies or even competing industries, as occurred when the *Brickbuilder* wrote in 1904 of the failure of "one of the standard brands of American Portland cement" at tests performed by investigators at the government's Watertown Arsenal. (*Cement Age* deflected this criticism by expressing doubts about the cement's typicality.)[85] Engineers understood this concern: a code of ethics issued by the American Institute of Electrical Engineers in 1912 contained a lengthy section on "Ownership of Engineering Records and Data." Here the society made clear that while certain innovations and work accomplished by an engineer for a client remained the property of the engineer, "if an engineer uses information which is not common knowledge or public property, but which he obtains from a client or employer, the results in the forms of plans, designs or other records, should not be regarded as his property, but the property of his client or employer."[86] Scientific findings in the context of commercial work did not enter some unbounded realm of universal knowledge but followed the principles of ownership one might see in material exchanges.

A second solution to the problems of scientific fixity involved redefining good scientific practice in a way that benefited those subject to its assessments. A good example of such efforts was the work of the National Association of Cement Users, which issued its own recommendations for concrete construction soon after its founding in 1904. A primary function of trade organizations was to position their members advantageously in the marketplace as regulations tightened competitive conditions. Charles C. Brown, editor of the journal *Municipal Engineer,* started a group to discuss the growing field of cement block manufacturing. The Engineering Congress of the 1904 Louisiana Purchase Exposition hosted an informal meeting to test interest in extending the scope of the organization to cover all aspects of cement. Response was enthusiastic: at a convention in Indianapolis the following year, Brown and his colleagues formed the National Association of Cement Users (NACU).[87]

The NACU began by appointing committees on concrete blocks and cement products; street, sidewalks, and floors; reinforced concrete; art and architecture; testing of cement and cement products; machinery for cement users; fireproofing and insurance; and laws and ordinances. The committees prepared standards and recommended practices on these subjects. Initially members expressed the greatest interest in concrete block and sidewalk production, but by 1908 reinforced-concrete construction had garnered a great deal of attention as well. Tilt-up construction, a particularly quick and inexpensive means of building with reinforced concrete, became a significant aspect of the NACU's investigations.[88]

An insight into the NACU's self-definition can be found in its leaders' belief that they were constantly faced with the "problem of trying to meet demands of both theorists and practical constructors."[89] Accordingly, many aspects of the NACU's recommended code were more liberal than municipal codes. Similarly, while it adapted a good portion of the ASTM-led joint committee's 1909 code for cement and concrete, the association pinpointed aspects of that code that it found to be too conservative. The NACU's leaders believed that the formulators of most codes for concrete "tended to penalize the economy of the medium in their attempt to eliminate chaotic conditions" and standardize practice. Of particular concern to them were the joint committee's recommended allowable stresses for reinforcing steel and moment factors for floor slabs. NACU members proposed more liberal allowances for both.

In 1916, still at odds with the joint committee's "rigid" and "arbitrary" rulings, the National Association of Cement Users—by now renamed the American Concrete Institute—promoted the idea that instead of including definite values for allowable stresses in reinforced concrete, codes should express appropriate stresses as a percentage of the twenty-eight-day compressive strength of concrete mixes. This innovation could be seen simply as the expression of the NACU's confidence in new concretes proven to reach a strength of 3,000–3,300 pounds per square inch (as opposed to the 2,000-psi concretes of the previous decade). However, it certainly also left room for materials suppliers to provide lesser products; a concrete mixture might be accepted at a percentage of its ultimate strength but never reach that strength. Suppliers might take this tack not out of disregard for a building's safety but in the belief that code requirements for concrete strength were too rigorous.

A third way in which manufacturers might evade problems of fixity echoed university and commercial efforts to cast the character of "inappropriate" testing personnel in doubt—attaching the potential utility of a scientific procedure to the competencies of its user, rather than to the procedure itself. Cement producers believed that the assurance or denial of quality that testing offered could be misassigned, and they drew attention to this type of malpractice. As one member of the ASTM wrote in 1907, "The inspecting laboratory, too, has the same problem to deal with that mill representatives have in regard to the ignorance of users of cement who often have crude ideas of cement testing and who make crude or field tests of cement. . . . These persons will imagine something wrong with good cement. Their crude tests will perhaps confirm their ideas."[90] Of course, this approach held some dangers for manufacturers. These formulations reinforced the belief that only university-trained experts should have responsibility for performing tests, a conception that could actually diminish manufacturers' control over the use of science in production matters. In addition to their worries about ignorant testers, cement producers and testers also feared that intentional usurpation of scientific authority posed a threat to "honest competition." Ernest McCready, the general manager of a commercial testing laboratory, wrote in 1907 that when a construction engineer arrived at his own test result and solicited bids from cement suppliers based on that result, as was common practice, unscrupulous suppliers placing bids might well do anything to achieve that same result. He warned that "it is 'up to' the manufacturer. He must prove by his own test at least, that his cement will do as it is told if he expects to bid under these specifications. So, nothing matters but the results. No mention is made of the methods or conditions under which the tests are made."[91] McCready and others believed that cement manufacturers might employ whatever means were necessary to yield desirable test results. For example, they might claim "normal consistency" to be very wet or very dry, depending on which offered easier handling. The idea that scientific authority might disengage itself from reliable, disinterested parties and come to rest on the "undeserving" could hardly be guaranteed to work in the manufacturers' favor.

Taken together, these various accounts and cautions demonstrate that the testing of building materials could bring to products the imprimatur of science even where actual scientific procedure was absent. That imprimatur was so powerful that the reputations of commercial products to which it adhered could be made or broken. The institution of uniform testing procedures, a priority of authors of model specifications (and described in chapter 2), appeared to many people involved in the science and business of building materials to be a means of preventing false crediting or discrediting of materials. In other words, uniform testing was intended as a preventive not just against building collapse, or against faulty practice within the field of materials expertise, but more broadly against the fraudulent use of scientific authority, a significant danger to products and professions in an industry that was coming to rely heavily on scientific and technological knowledge. We can recognize that uniformity was perpetually subject to subversion, but that even its invocation may have had a significant social function in granting, or denying, occupational influence.

Conclusion

As we observe the use of science for commercial advantage and find rampant anxieties about tests misused and results misassigned, it is tempting to lapse into questions of honesty. Were the American Society for Testing Materials, the Portland Cement Association, and building businesses in general operating in good faith as they engaged in science for profit? Wrongdoing was not unheard of. The U.S. Army Corps of Engineers warned consumers of cement in 1901 that the mere presence of testing procedures in a cement plant was not a guarantee of quality. If cement manufacturers claimed unusually high tensile-strength test results for their products, it could be due to deliberately or otherwise faulty testing methods or to product adulteration. Excess lime in a cement, which could be temporarily masked by the addition of sulfate of lime, would yield high early strength in a mix that would ultimately weaken.

The corps engineers also warned consumers against the practice of offering cash bonuses for cements testing above a fixed high point. Some purchasers believed that cement manufacturers would be more likely to produce higher-grade cements if eligible for a bonus to be awarded on the basis of specific gravity, soundness, and fineness tests. Through this practice purchasers could buy their cement from the lowest bidder but could, at a little extra expense, "induce and foster competition" among manufacturers and "practically eliminate the necessity of rejecting cements." At least a few manufacturers were amenable to this suggestion, and it reappeared throughout trade literature from 1900 to 1920. The corps engineers warned, however, that "cements so obtained are likely to be unsound in a manner not easily detected in the time usually available in testing."[92]

Similarly, codes of ethics for architectural and engineering trades commonly warned that their members should avoid underbidding, bribes, and conflicts of interest arising from serving multiple clients—suggesting that these did occur to some degree.[93] But to pose questions about the ethical intentions of builders may not be entirely fruitful. As noted, it is always difficult for the historian to discern whether wrongdoing has been deliberate or not. But more important, we can understand that science is exactly what it appears to be here: an enterprise that has a fluid character based on its social efficacies. Leading members of the ASTM defined what they thought to be "good science." They believed that equitable commercial relations would be assured wherever that good science was practiced. University professors, members of the Portland Cement Association or the National Association of Cement Users and other participants in the building industry tweaked or countered that definition. The actual incidence of corruption in the concrete construction industry is secondary to the idea that technical expertise offered a powerful source of economic and occupational influence. Few people in this arena pictured some ideal technical practice beyond the reach of commercial or disciplinary self-interest; rather, they saw fairness in the imposition of their own technical ideals.

The changing definitions of reliable scientific practice, and the variable and ingenious manner in which experts and businesses established the trust of colleagues, point us to the compelling force of technical expertise engaged for social purposes. Nonetheless, if many of the interactions recorded in this chapter were adversarial, they reflect a sphere in which almost every player had some measurable degree of choice in how his work was undertaken. On the scale of occupational status, engineers, contractors, and even materials suppliers were relatively elite, autonomous individuals. The use of systematized technical knowledge among these groups offered each party at least the possibility, if not always the attainment, of personal or corporate gain. We can turn now to a set of notably less equitable relations configured by the developing technologies of concrete: those between the managers and owners of engineering firms that built concrete structures and the vast body of employees who performed the physical work of building.

CHAPTER FOUR

The Business of Building
Technological and Managerial Techniques in Concrete Construction

> As in battle, so in building, where the officers go the men will follow.
> —*Aberthaw Construction Company, 1920*

We have been regarding the production of knowledge about concrete in the early twentieth century as a task shared—willingly and otherwise—by university instructors, their students, materials producers, building trades, and lawmakers. We have seen how written representations of this knowledge determined the bench-top techniques of college students and the commercial exchanges of entire industries, and how professors, young testing engineers, cement makers, and contractors came together around this body of knowledge to vie for occupational advantage and business profit. We can now turn to the final physical and administrative application of this new information, science-based technique, and commercial ingenuity: the erection of concrete buildings between 1900 and 1930, particularly the massive utilitarian factory buildings for which concrete was most frequently employed in the United States. The flow of knowledge and information about concrete does not diminish in social import here; rather, it is associated with a further set of material, fiscal, and occupational challenges.

From its first large-scale use at the turn of the century, concrete has always been defined by its promoters and historians as special—different from all other building materials because it is "manufactured" on the construction site rather than assembled.[1] Whether concrete is truly unique in this regard depends on what definition of manufacturing one uses. One could say that a steel building frame is manufactured on site because it does not achieve its structural character until beams and girders are riveted or welded together. Surely from the standpoint of quality control an overheated rivet, or for that matter a poorly shaped marble block or badly driven nail, is not categorically different from an overdiluted concrete mix. But however unique—or not—we decide concrete to be among building mediums, the formulation is a telling one. It raises the specter of concrete as a material that cannot be left to the exigencies of ordinary building methods or ordinary personnel. It indicates that promoters wanted to link the use of their product with special skills, maintaining the call for expert intervention from first handling to end use in the field. The idea that concrete buildings are manufactured rather than "erected" also suggests that techniques of modern mass production might serve very well any firm trying to construct such structures profitably.

There are some familiar patterns here. Purveyors of concrete buildings sought to resolve the tension inherent in providing expertise in commodified form while holding on to a distinct and marketable identity for that expertise. As materials experts had worried regarding the formation of standards and specifications for concrete, builders knew that the economies of scale inherent in their construction processes might well make proprietary control difficult. Concrete construction celebrated standardization of process and product. Well-capitalized engineering firms, pursuing cutting-edge managerial techniques, organized motions of workers and the handling of materials with tremendous precision. Such firms established elaborate bureaucratic structures for the direction of their workforces. The planning, design, and erection of buildings; tasks of hiring, accounting, and maintaining inventory; and many other features of construction were all subject to careful study and deliberate arrangement. This standardization of work, both manual and administrative, in factory construction was paired with standardization of product: thousands of concrete factories erected after 1900 were almost indistinguishable in appearance. Their exposed concrete skeletons, uniformly large windows, and complete lack of exterior ornamentation rendered them virtually anonymous as they filled more and more of the industrial landscape in America. The proliferation of prefabricated standardized supplies and parts for concrete buildings—steel reinforcing, metal-framed windows, and many other architectural elements—followed from and further encouraged this design trend. But as was the case in the development of standards and specifications, the routinized procedures for erecting concrete buildings were firmly attached to their promoters—not to be adopted by an infinitely wide array of practitioners. The building firms used slightly different techniques for securing jurisdictional control of the new technologies, but they were no less successful than the materials specialists, and we will see that the two efforts were related in some important ways.

How did the builders offering this streamlined, standardized construction system manage to identify themselves so closely with the technology? First, they crafted the knowledge involved in concrete construction to resemble the managerial approaches being used by their clientele—most often successful owners or directors of manufacturing, processing, or distribution enterprises. The self-descriptions of factory-building firms cast their expertise in familiar terms but then clarified that it was similar in form to that possessed by most building buyers while not identical in content and, most crucially, *not transferable* among even modern, high-level occupations.

Second, factory-building firms promoted their own enterprises as engines for social and fiscal change of benefit to their clientele. The large, modernized building concerns displaced not only old technologies but also existing patterns of economic opportunity. As the big new firms expanded their influence among factory buyers, local networks of small contractors and suppliers lost significant commercial leverage, and in an even more sweeping change, skilled building trades found themselves at an unprecedented disadvantage. Bricklayers, stonemasons, carpenters, and plasterers in every American city had established large and self-perpetuating communities, and in many locales they had consolidated into chapters of international unions by the 1880s. Configured by apprenticeship and patronage systems, these groups fit poorly with the control- and profit-driven plans of industrialists seeking to erect new physical plants at the greatest possible speed and least cost. We must be cautious in accepting at face value the accusations of corruption and self-interest that building buyers hurled at small local builders and craft groups, but we can certainly recognize a set of conflicts that the new integrated engineering firm selling concrete factories promised to dispel.

If we scrutinize the two strategies developed by modern building firms to attract commercial clients, an intriguing irony emerges. The firms saw their technical knowledge and management techniques as integrated and comprehensive. In this way the skills they advertised as their own closely resembled in form, if not content, the very bodies of knowledge they sought to displace: the artisanal talents of the small, apprentice-based local building concerns. This retrograde character for new building methods resonates with the blended nature of science-based materials testing—bearing highly "modern" systematized features and extremely ill-defined "artlike" elements.[2] A picture of a supremely self-conscious and ambitious stratum of business people emerges, twin in many ways to the elite of cement and concrete experts on whose knowledge they drew.[3]

This chapter describes the work of firms that built concrete factories, offering first a general history of factory construction and then an account of the increasing efficiency of concrete construction and the popularity of concrete factory buildings after 1900. A case study of one very successful engineering firm that specialized in factory construction follows. The technical and social programs of factory builders are linked throughout, challenging traditional explanations of the role of technology in shaping work practices. In his landmark 1930 study of the American building trades, William Haber pinpointed mechanization and other technological changes in the construction industry as *causes* of the industry's early-twentieth-century managerial and administrative reorganization.[4] In this same period, however, businesses throughout the country were pursuing similar reorganization in contexts ranging from heavy-manufacturing plants to clerical offices that used very little technology. Further, at least one inventor of a streamlined construction process referred to his products as "ideally suited for scientific management of the construction site," implying that the new management methods were already in place.[5] Taking this larger perspective, in which technological changes do not necessarily precede administrative changes, the causality of Haber's argument comes into question. It would appear instead that the managerial and technological changes in the construction business after 1900 could have been mutually causative, and that they reflect a combination of logistical and social agendas.

History of Factory Design and Construction

The forms and construction methods applied to factory buildings in the United States between 1900 and 1930 resulted from a steady growth and formal refinement in industrial architecture.[6] Before 1800, manufacturing in the United States generally took place on a small scale in wooden structures very much like sheds or barns. Rural and urban workshops usually employed only a few workers and did not require specialized structures.[7] After 1800, manufacturing in many cases shifted from the use of a few independent machines to series of larger, interconnected machines. A single mill, for example, might perform both spinning and weaving. At first only the attics of conventional buildings could accommodate such series of machines; roofs were supported by wide trusses that created uninterrupted expanses of floor space below. But within a decade or two buildings were being designed to accommodate the new processes and larger numbers of workers.

By the 1830s, purpose-built structures, particularly for textile mills, were common. Their design accommodated the shafting systems that transmitted water power throughout the facilities. The metal shafting common at this time allowed effective transmission of power to distances of about 100 feet, and because shafting was most often based on a system of vertical shafts and bevel gears, many of the mills of the time were tall, narrow buildings.[8]

Different builders preferred different construction technologies for their factories, depending on regional tastes and the intended use of the building. Larger mills after 1800 were likely to have timber skeletons with brick or masonry walls. This combination produced a sturdy structure with small windows penetrating heavy exterior walls. After 1810 some larger industrial buildings following European precedents were constructed in part of cast iron. Cast-iron ribs with copper sheeting replaced traditional timber roofs, and cast-iron columns replaced timber posts. Cast-iron columns could support wider spans than conventional timber framing, a fact of importance to industries using larger machines. In many instances, iron members supported brick vaults or vault-shaped iron plates, which were then filled to floor level with concrete. This technology could easily be used to build a seven- or eight-story building.

Despite such advantages, American industrialists did not embrace iron construction wholeheartedly.[9] In much of the Northeast and Midwest, large timbers were easily obtained at this time, so many industrialists relied on an adaptation of timber-skeleton construction that used very heavy members. This technology came to be known as mill construction or slow-burning construction. Developed for New England mill companies after about 1825, it was based on what one historian has called "deliberate over-design."[10] Large timber members would char but not burn, allowing the evacuation of occupants and goods in case of fire. Connections between members were made in such a way that if one element broke or fell in a fire, others would not be pulled down with it. Small members or spaces in which fire could flourish were avoided. The technology was endorsed by the powerful New England mutual fire insurance companies and gained a wide following around the country.[11]

Mill construction offered larger interior bays than traditional timber designs because the broad timber supports could stand farther apart than narrower ones. Exterior walls could hold larger windows for the same reason. It was easy to provide openings for shafting in the wooden floors and to add or change fixtures. Even after the displacement of cast iron by rolled steel—a more flexible and fire-resistant structural material—around 1900, timbered mill construction retained many adherents.

The large dimensions of timbers in slow-burning factory buildings presaged the reinforced-concrete skeletons of early-twentieth-century buildings. The wooden beams frequently measured 14 inches on a side. (When steel framing became a viable replacement for iron in large buildings, it was often clad with fire-resistant ceramic materials until it too reached these trusted dimensions.) But other developments also helped set the stage for design trends to appear in post-1900 factory buildings. Over the course of the nineteenth century, factory buildings became increasingly standardized in construction and appearance. Many mills of the early and mid-nineteenth century had featured gables, turrets, or distinctive towers that made the buildings similar in contour to large homes or churches. But as developers of mill sites in Lowell, Massachusetts, and other eastern cities established markets in commercial real estate, it was to their financial advantage to promote a more standardized type of manufacturing plant. Idiosyncratic design could lead to unpredictable erection and maintenance costs and increased fire risk. A simplified floor plan also meant that a building might be convertible to different uses for future owners. Certainly, many industries required highly specialized structures for the refinement or handling of materials during manufacturing.[12] However, as the nineteenth century progressed, many industries and real estate speculators followed the textile industry's lead and adopted uniform mill structures.[13]

The Standardized Factory Building

A contributing factor in this standardizing trend was the practice of textile machinery firms, and later of fire insurance companies and steel suppliers, of issuing free building plans to purchasers of their products or services. This practice eventually lost favor as factory design and construction became the responsibility of dedicated experts in the field, but a pattern of planned, standardized physical plants had been initiated. Uniform structures could be built at predictable cost and could provide the owners with the benefits of other builders' experience.

Factory owners were also approaching plant layout with greater foresight. Industrialists and engineers believed that the processes that were to occur within a factory should determine to a large degree the nature of that factory's construction. By 1900 a few types of building designs had emerged to suit what were thought of as general classes of manufacturing processes.[14] These planned factory buildings took into account manufacturing processes, relations among these processes, requirements for natural lighting, systems of power transmission, arrangements for shipping and receiving (such as rail connections), storage needs, sewage and drainage conditions, and lot size. Although some manufacturing processes called for factories of different shapes, such as L, T, U, or H configurations, industrialists and builders considered rectangular buildings to be the most economical and versatile.[15]

A few distinct types of rectangular buildings began to fulfill numerous manufacturing functions. The three most common types of factory in 1900 were the one-story general utility building; the one-story building with sawtooth roof; and the multistory factory building, of either light or heavy design. The one-story general utility building was best suited for foundries, forges, and heavy-machine shops that required unbroken expanses of floors and high ceilings. These buildings might feature open sides and heavy-duty cranes that could move materials around the space. Windows could effectively light such a building up to 60 feet wide. Beyond that width, additional lighting could be provided overhead by sawtooth roof monitors. A one-story building with a sawtooth roof was likely to be a lower structure in which finer assembly operations, such as engine manufacture, took place.[16]

The third dominant form of industrial building in 1900 was the multistory factory that carried forward the mill traditions of the preceding century. Such buildings could accommodate light or heavy manufacturing and were chosen by firms that made products ranging from trucks to tooth brushes to breakfast cereal. Brick and heavy timber were the most familiar materials for these multistory factory buildings, but steel was also recognized as a reliable framing material. If clad in plaster, tile, terra cotta, or concrete, steel members gained a substantial degree of fire resistance. Steel trusses offered the possibility of wide roof spans as well. After 1900, however, it became increasingly common for builders to select reinforced concrete as the primary material for factory construction (fig. 4.1).

Through the first three decades of the twentieth century, two basic forms of construction were employed for reinforced-concrete factory buildings (fig. 4.2). The beam-and-girder method most directly imitated traditional timber-framed construction by relying on a system of columns, cross beams, girders, and slab floors. The flat-slab method, based on innovations of Robert Maillart in Europe and C. A. P. Turner in the United States, offered a simpler form. It eliminated beams and girders in favor of a flared-top, or "mushroom," column and slab floors with more substantial reinforcing.[17] The absence of beams and girders left more headroom—of particular value in rental properties where charges were based on usable space—and permitted light to diffuse more completely throughout a work space. This spare headroom also somewhat eased the problem of adding shafting or fixtures to a completed concrete structure. Flat-slab construction also required less carpentry for formwork, and was thus less expensive than beam-and-girder work. By 1920 it was clearly the preferred construction method for large reinforced-concrete industrial buildings.[18]

Concrete was a somewhat unfamiliar primary material for large buildings, but it offered a number of fairly obvious advantages, including the availability of raw materials. A major steel shortage occurred in the United States in 1897 and 1898. The materials required for reinforced-concrete construction were not subject to such shortages. Sand and aggregate were always available, often from local sources, and the production of cement in this country grew from some 8 million barrels per year in 1900 to 35 million in 1905.[19] Further, reinforced concrete offered a means of framing that allowed minimal infill between exterior columns, and therefore a greater proportion of exterior walls could be devoted to windows. The reinforced-concrete-framed factory building thus gained the nickname "daylight factory." Reinforced-concrete construction was also relatively resistant to vibration caused by machinery or by earthquakes, presenting advantages of safety and convenience. Concrete floors were easy to seal against dust and were therefore more sanitary than wood or brick surfaces—an advantage in food processing or pharmaceutical manufacture.[20]

Perhaps the advantage most often cited for reinforced-concrete factory buildings by builders and developers in the early twentieth century was the material's great fire resistance. As dramatically demonstrated in a 1902 fire

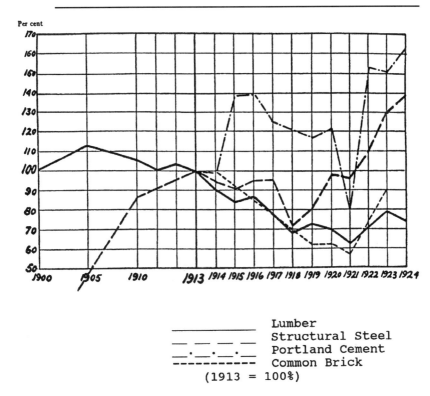

FIGURE 4.1. Graph produced by the National Lumber Association showing the rising popularity of concrete among major building materials after 1900. *From William Haber, Industrial Relations in the Building Industry (1930; reprinted New York: Arno and New York Times, 1971)*

at the Bayonne, New Jersey, Pacific Coast Borax plant, concrete buildings could emerge virtually unscathed from fires that would crumple steel. Such durability had immediate practical appeal to factory owners and the indirect advantage of carrying with it lower insurance premiums. At least one architectural historian has claimed that it was reinforced concrete's superior fire endurance that ultimately made it a more popular choice than slow-burning timber or steel-frame factory construction.[21] However, this explanation oversimplifies the reasons for the popularity of this building type.

First, as one prominent engineer noted in 1911, timber construction could perform as well as reinforced concrete in a fire: "A properly designed building of mill construction, if protected with sprinklers, fire-fighting appara-

FIGURE 4.2. Favorable comparison of lighting conditions obtained using the flat-slab-ceiling method of reinforced-concrete construction (*top*) with the older, beam-and-girder method, which more closely resembled traditional timber mill design. *From Willard L. Case,* The Factory Buildings *(New York: Industrial Extension Institute, 1919)*

tus, and cut-off walls is in many cases as reliable as industrial conditions demand."[22] This argument has relevance because even in 1910, with the production of cement in the United States increasing, reinforced concrete was a more expensive method of building than timber mill construction. If mill construction was both less expensive than concrete and potentially of equal fire resistance, the ascendancy of the reinforced-concrete factory after 1900 cannot be attributed solely to its performance in fire.[23]

If we look beyond fire resistance as the primary reason for reinforced concrete's appeal to factory owners, we can begin to appreciate another feature of these buildings: their remarkable uniformity, which derived from the conditions of their erection as well as from emerging notions of architectural respectability. Many reinforced-concrete factory buildings shared the same shape, methods of construction, and approach to ornamentation to a degree not found in office buildings, hospitals, schools, or other utilitarian buildings—not to mention residences and civil buildings—of the day. Catalogs of factory builders and trade literature show what is clearly a style of building, varying in size but in few other ways. The typical reinforced-concrete building erected between 1900 and 1930 was rectangular, usually with concrete skeletons unhidden by brick cladding or ornamentation. Occasionally the simple contours of such buildings were interrupted by a tower that housed stairways or bathrooms, but only rarely. Where ornamentation was used, it was usually in the form of a simple cornice or of the beveled edges of columns, beams, or sills that could help prevent crumbling, or "spalling." The company name or the function of a building, such as "Bottling House" or "Machine Shop," was sometimes cast directly into a facade, but in general these buildings were constructed with few features that might assert their identity.[24]

The buildings also had an internal uniformity: plans were drawn up for a single bay that could be multiplied many times to create a factory of desired dimensions. This repetition minimized design expenses and helped builders and building owners estimate construction costs. The scheme of repeating bays also meant that windows were uniform in size, enabling builders to take advantage of the growing selection of standardized, mass-produced metal or wooden window units.

In their uniformity, reinforced-concrete factory buildings of the period 1900–1930 embodied characteristics of any mass-produced artifact of the day. They owe their existence to the application of modern production and business methods to building. The standardized reinforced-concrete factory building in the United States was a commodity that was successfully mass-produced and marketed, and its history is a history of those activities with all their consequences for professional and labor groups. The uniformity of this identical "product" had a social significance that went hand in hand with its more explicit functional advantages.

Proliferation of the Reinforced-Concrete Factory Building

Reinforced-concrete factory buildings in the United States were first promoted by a handful of accomplished and ambitious practitioners. Ernest Ransome, an engineer known for his innovative building designs, improved concrete-handling machinery, and patented reinforcement system, is a particularly significant figure.[25] He developed a simplified version of François Hennebique's reinforcement using a twisted square iron rod. His industrial plants of the late 1880s and 1890s were widely lauded, and his 1903 United Shoe Machinery Plant, of Beverly, Massachusetts, was the largest reinforced-concrete industrial plant to date. It featured precast beams set into slots in column heads, as well as immense expanses of windows. Ransome's utilitarian factory designs broke with traditional decorative forms; Reyner Banham writes that the United Shoe plant's austere exterior was "entirely admirable in its appropriate mixture of decorum and puritanism."[26]

Albert Kahn is also often cited in the history of reinforced-concrete factory architecture. Using a system of reinforcing developed by his brother Julius Kahn, Albert Kahn designed many structures for the Packard and Ford automobile companies between 1900 and 1920. Although Kahn's residential and civic commissions remained elaborate and eclectic, he achieved what clients and architectural critics considered to be appropriately austere and economical designs for industry.[27] Other architectural firms that achieved celebrity for their industrial commissions in these years include Purcell and Elmslie; Pond and Pond; and Schmidt, Garden, and Martin.

However, reinforced-concrete factories were often built without the involvement of well-known architects, or of any architect at all. The buildings' proliferation on the American urban landscape may have grown from the work of known figures, but the vast majority of these buildings were designed and erected anonymously, within a world of routine commercial transactions. Lesser-known firms learned of new technologies and designs through trade publications and professional organizations and through patents taken out by leading designers. These firms offered their services to factory owners who could not afford the best-known builders, disseminating the reinforced-concrete factory building to industrial districts around the country.

With the exception of some mill-building specialists, American construction firms prior to 1900 were small and relied on transient labor forces. The introduction of new concrete building technologies and new systems of management after 1900 was concurrent with the formation of a new type of firm. These firms offered services in a manner unprecedented in the construction industry.

In the first decades of the twentieth century, an industrialist wishing to erect a new facility for his business had in general three choices. First, he could employ his own forces for all construction work. He would in this case enlist an engineer or architect to draw up plans, hire subcontractors for specialized work, and assume all responsibilities for erecting a plant. This was a procedure quite common in the nineteenth century, but by 1910 it had become favored only in cases in which an industry might require a construction force to remain on hand to operate a facility, such as one designed for complex distilling or refining processes. The number of subcontractors required to erect a typical factory with powerhouse and office building was so great—ranging from excavation forces to electricians to specialists in chimney and flue construction—that most factory owners were reluctant to take on the job unaided.[28]

A second option involved the owner's soliciting plans and specifications for a factory building from an engineering firm and then submitting them to prospective building concerns or general contractors for bids. This "letting by contract" could entail the enlistment of a few contractors or a great many. The engineering firm would coordinate the work of the selected contractors. As noted in chapter 3, the question of whether engineer or contractor ultimately determined the exact products and service providers used on a project was under constant debate, but this approach was substantially easier on the owner than taking on supervisory tasks himself.

However, a third option showed the greatest increase in popularity among factory owners at this time. This was the hiring of firms that included an engineering division able to design factory buildings and a construction division able to erect the buildings from start to finish. Such firms usually maintained separate departments for advertising, drafting, estimating, accounting, purchasing, expediting, and construction. With these facilities a building firm could select the best site for a client after having its own staff study local geographical, supply, and labor conditions, and then coordinate every aspect of construction from excavation to final painting.

A number of the engineering firms that operated along these lines were very successful. Perhaps best known today are the Lockwood, Greene Company and Stone and Webster.[29] Lockwood, Greene was founded in New England in the early nineteenth century. Although the firm achieved a specialization in the management of textile mills, its factory commissions included manufacturing plants for products ranging from fertilizers to pianos. Stone and Webster, founded in Massachusetts in 1889, built immense industrial facilities, often using flat-slab reinforced-concrete construction. In the early twentieth century the firm was a leader in the construction of power plants, but it also built complex refineries for U.S. Rubber, American Sugar, and other large concerns. It erected factory buildings of smaller scale as well, sometimes also providing manufacturing or power-supply machinery used in the plants.[30] Other firms of slightly smaller size performed similar services on a regional basis. These were particularly common in the Midwest and Northeast, where manufacturing industries experienced substantial growth between 1900 and 1930.

Their functional departmentalization makes the engineering/building firms kin to other mass-production industries of the day. The firms' appeal for factory owners was based on the quality of services and savings they offered as well as on a shared ideology. Not only were the complexities of dealing with bids and subcontractors eliminated for factory owners who turned to the modern building firms, but the costs added as each contractor and subcontractor sought profit were also removed. In his 1930 report on American construction trades, William Haber summarized other advantages that the integrated engineering/building firm held for owners:

> Conducted on this basis, building construction can be carried on under the best methods, taking advantage of the latest improvements in machinery.... [The large company] has a special purchasing department whose particular task it is to buy all materials and through its expediting division to follow up all deliveries so that delays may be minimized. The smaller contractor leaves this important task to the superintendent, who must also attend to other duties, such as those connected with production schedules and labor problems.[31]

Such unification and centralization allowed the multifunction construction company to exploit economies of scale and the emerging art of coordinating production tasks.

New Technologies of Reinforced-Concrete Construction

We can divide the strategic operations of factory-building firms in the early twentieth century into activities of two kinds. First, firm owners made technological choices from a variety of specialized products and services. In general, the physical means of building that they chose became progressively more streamlined, standardized, and mechanized after 1900. Second, building firms selected certain administrative methods for building projects. The managerial techniques they preferred for the new construction processes tended towards centralized direction of dispersed field operations. The two categories of decisions were related. New equipment often altered workplace organization, while builders' desires to employ a little-trained workforce created a market for appropriate new technologies. The technological and managerial choices surrounding reinforced-concrete construction reflected a range of strategies for increased production and workplace control. In pursuing these ends, managers of specialized building firms were making a sophisticated and conscious effort to establish the multifunction construction company as a modern business enterprise.

The technological procedures involved in the erection of a reinforced-concrete building after 1900 may be divided into site preparation (by excavation or other means); creation of foundations for walls and columns; erection of wooden forms; placing of iron or steel reinforcement in those forms; mixing and pouring of concrete; removal of forms after the concrete has set; finishing of exposed surfaces; and installation of doors, windows, roof coverings, and sprinkler systems and other plumbing. In many cases a concrete-mixing plant would be erected while the foundation was being laid. Initial deliveries of materials and the erection of a carpentry shop for the preparation of forms could also take place at this time. The drive to hasten and economize industrial construction addressed processes and the flow of materials at each of these junctures.

As we have already seen in the work of technical experts involved in concrete testing, quality control for concrete involved operations of two types. On the one hand, as a pourable medium, it could be handled efficiently on

142   REINFORCED CONCRETE

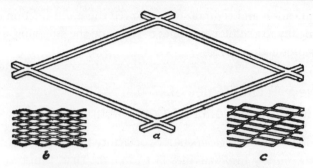

Fig. 57.   Expanded Metal of Standard Mesh.

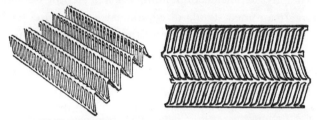

Fig. 58. A Special Form of Expanded Metal Reinforcement.
Fig. 59.   Expanded Steel Lath.

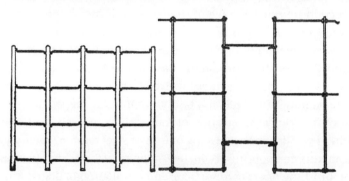

Fig. 60   Rib Metal.   Fig 61.   Wire Fabric and Clamp
Mesh and Fabric Materials for Reinforcement.

FIGURE 4.3. Selection of prefabricated reinforcement for slab concrete as employed in walls, floors, and ceilings. Such systems, intended for sale to concrete construction firms, removed the work of reinforcement assembly from the construction site and largely eliminated the use of skilled fabricators in favor of machine-based production. *From William A. Radford,* Cement and How to Use It *(Chicago: Radford Architectural Company, 1919)*

a mass scale. In theory, forms once erected could be filled without interruption, and on a well-organized project, pouring could continue on one portion of a building while another portion set. On the other hand, the erection of wooden forms and the placement of reinforcement could require slow, precise attention from costly skilled workers. Managers of construction enterprises systematically sought to translate the second type of operation into the first. For example, as one engineer summarized in 1906, the essence of economy in concrete was to be found in the duplication of forms and the elimination of architectural details that complicate form construction.[32] Concrete construction was greatly expedited when, after 1900, outside suppliers increasingly took over the construction of forms and the assembly of reinforcing rods. These auxiliary businesses, located off the construction site, mass-produced materials that otherwise had to be individually fabricated in the course of building. Some intricate types of forms and reinforcement continued to be fabricated by workmen on the building site, but enough were standardized and mass-produced to effect substantial economies.

A particular type of design, such as Turner's mushroom column, often became known as a "system." Commercially produced reinforcement systems first appeared in the 1890s. They capitalized on the idea that reinforcement could be bent and assembled by machine in quantity before reaching the construction site. In 1906 *Cement Age* published a review of ten commercial systems of reinforcement; by 1914 dozens of firms were advertising systems of preassembled reinforcement in *Sweet's Catalog*. These advertisers ranged from prominent building firms that produced reinforcement as a sideline to specialized manufacturers. The Standard Concrete-Steel Company offered "Consulting Engineers and General Contractors for 'Standard' Systems of Reinforced Concrete Construction."[33]

Among the best known of these product lines was the Kahn System, developed by Julius Kahn. Kahn's system of "trussed steel bars" featured rolled-steel bars of diamond cross-section with bent-up "wings" attached to either side. The wings countered the shearing forces found in concrete beams, adding from 20 to 30 percent to the strength of a beam (fig. 4.3). The Kahn Company also produced spiral hooping for columns and expanded metal, a sort of metal netting for reinforcing flat slabs or walls. (Other companies offered thin corrugated steel sheets for this purpose.)[34]

Some reinforcement makers were clearly of lesser sophistication than the large concerns. The Hinchman-Renton System offered reinforcing made from "ordinary barbed wire."[35] But even the simplest product lines offered purchasers economies of scale based on replacing individually assembled reinforcement with mass-produced assemblies. For example, in 1906 the Clinton System featured "wire cloth," electrically welded metal fabric produced in 300-foot rolls.[36] Some firms offered "unit girder frames" that constituted preassembled reinforcement for entire beams or girders, ready to be set into place by three or four relatively unskilled workers. The Unit Concrete Steel Frame Company went so far as to provide sockets that would be fitted into the bottom of forms to assure the correct placement of the reinforcement unit.[37] The costs of purchasing this kind of fabricated reinforcement were offset by savings in labor on the construction site and by the prevention of excessive steel use. Mass-produced steel reinforcement had become so affordable by the early 1910s that even large construction firms stopped fabricating their own reinforcing rods.[38]

In addition to purchasing preassembled reinforcing, by 1903 builders could make use of precast concrete elements, avoiding the difficulties of erecting forms and of pouring concrete above ground level. Elements were cast on the ground or in workshops with reinforcement in place and then assembled when cured. Generally a light slab was poured in place to unite the assembled frame elements. For the first decades of the century, this "unit construction" involved casting relatively small pieces on the building site. Ernest Ransome pioneered some of the most ambitious uses of the technique and referred to his method as "monolithic unit construction" to convey its structural similarity to conventional methods of concrete construction. In the Ransome Unit System, columns, beams, and girders were cast in an empty lot next to the building under construction and hoisted into place. For this work Ransome employed "gang molds," in which several like pieces could be poured simultaneously (fig. 4.4). By 1911 he could claim that his system was "10 per cent lower in cost than monolithic" and that it was "easier, quicker, requires less skilled labor and is more exact and cleaner." It also permitted concrete construction to continue through the winter months because precast elements could be prepared in advance or inside heated sheds. By 1915, engineers had created systems that offered everything from walls to window sills.[39]

The notion of systematic procedures and sets of products for reinforced-concrete construction pervaded the industry, which had developed a penchant for products that promised simplicity of operation and predictable conditions on the building site. The economic advantages of system-based

FIGURE 4.4. Workers placing concrete beams by means of a "gang sling." This device and other rationalizing techniques, such as "gang molds," assured that beams need never be handled individually. Note the advertisement of Ernest Ransome's patented methods, largely developed to exploit such mass-production features of concrete construction. *From* Cement Age 12 *(March 1911)*

construction were intertwined with the commodified nature of new technological knowledge. Words like *system* and *unit* signaled the presence of specialized technical knowledge and rationalized production methods. Systems—whether of reinforcement or of precast elements—were subject to licensing, patenting, and other marketplace controls. François Hennebique established his program for licensing his system of concrete construction in France in the 1890s, and many other firms in Europe and North America capitalized on the reputation of the parent organization. For example, the Turner Construction Company advertised its status as a "licensed agent of the Ransome System."[40]

Hennebique had found licensing to be an effective way to expand the use of his system and increase his profits while keeping control of the quality of buildings created under its name. Patents served a similar function for some engineers. Kahn's addition of wings, or stirrups, to reinforcing rods, for example, was widely praised as a means of overcoming the shearing stresses that exerted destructive twisting forces on a beam. However, it could not be freely adopted by builders because it was patented. If a builder was unwilling or unable to pay for Kahn's patented prefabricated reinforcing, he could not employ the new technology.[41] Some concerns tried to turn their lack of property rights into an advantage. The John W. Allison Construction Company of Philadelphia claimed in 1907 that it offered clients greater flexibility and economy than building firms that "confined" themselves to the use of one type of bar or patented application.[42] However, claims to the rightful use of a new concrete technology were far more common on the part of construction firms competing for the business of factory owners.

Mechanization and the Flow of Concrete

In addition to introducing systems of reinforced-concrete construction, the building industry rapidly mechanized the concrete building site after 1900. Technologies for mixing and distributing concrete developed quickly. The production of cement had been greatly speeded in the 1890s by the introduction of the rotary kiln and other means for continually crushing, drying, roasting, and powdering cement ingredients. The mechanization of concrete mixing soon followed. Steam- and then gas-operated mixers proliferated between 1900 and 1920, steadily increasing the pace of concrete construction. The Ransome Company described its 1908 model in vivid terms: "The scoops might be compared to great shovels in the hands of a man powerful enough to handle them quickly."[43] By 1931, powered mixers could mix concrete batches of 56 cubic feet and achieve an hourly output of 80 cubic feet (fig. 4.5).[44]

The most important feature of all powered mixers was that they supplied a nonstop flow of concrete to waiting forms. They could be filled and emptied continuously, eliminating bottlenecks associated with hand-operated machines. New technologies arose for distributing this steady supply of concrete around the construction site. First, builders erected ramps and runways over which wheelbarrows of wet concrete could be transported. The wheelbarrows would be emptied into waiting forms. Systems of carts run on tracks replaced the ramps and wheelbarrows for large projects, and by 1910 there existed systems by which empty carts could be automatically returned to the concrete mixer for refilling.[45]

By about 1905, elaborate systems of hoists, towers, and chutes offered an

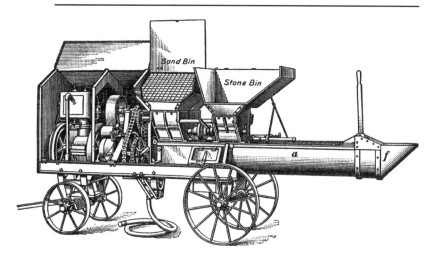

FIGURE 4.5. Portable concrete-mixing machine produced by the Eureka Company for small-scale commercial construction projects. Such machines became increasingly popular through the 1910s and 1920s because they permitted continuous preparation of raw materials and output of wet concrete and, importantly, allowed the proportions of cement, sand, gravel, and water to be adjusted without stopping the machine. *From C. K. Smoley,* Stone, Brick, and Concrete *(Scranton, Pa.: International Textbook Co., 1928)*

efficient means of distributing concrete on the largest sites. These apparatuses had superseded other means of distributing wet concrete by 1920 except in very intricate operations. Wet concrete was carried to the top of a tower by powered hoists and distributed by gravity through chutes covering areas as large as 800 feet in diameter. Flexible spouting facilitated these operations; in exceptionally large projects, towers were set up to move around a site on tracks. In many cases a few men would be required to spread the concrete evenly once it arrived at the form, but chuting and related methods of concrete delivery substantially reduced the time and labor costs of concrete construction.[46]

Most aspects of concrete construction were designed or redesigned between 1900 and 1920 to assure the incessant flow of wet concrete around the site. Machines that automatically opened and emptied sacks of cement appeared on the market, as did belt conveyors to carry the dry ingredients of concrete into mixing machines (fig. 4.6). Not surprisingly, the arrangement of these facilities on the construction site became a subject of study, as it too could affect the movement of materials. Clearly the size and shape

148 REINFORCED CONCRETE

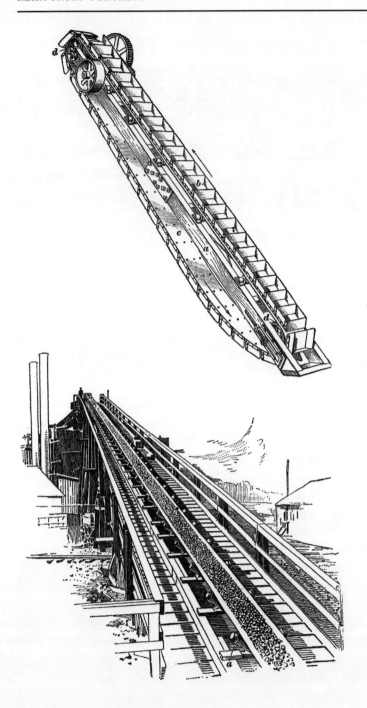

of a building determined something of the site's arrangement, but other factors were also significant. One construction company considered the cost of the concrete plant itself to be the most uncertain item in estimating the cost of construction. In advising building buyers of the variables involved, the firm's estimating engineer wrote that "the number and location of the mixers, towers and runs used on the job, layout and extent of storage space for aggregate, source and expense of power, etc., distance over which concrete machinery has to be transported, good or bad mechanical condition of rented machinery, rental rates of machinery . . . and many other variable expenses go to make up this cost."[47] Each step in the handling of concrete offered the risk of wasted time, wages, and materials.

Contemporary literature outlined approaches to overcoming these difficulties. Small projects required a few strategic decisions based on site layout and the possibility of moving materials with gravity. Because it could take considerable time and effort to move a concrete mixer and stocks of aggregate and sand, a location for these items would be selected that either was centralized or allowed the shortest possible length of chuting to the place where the greatest amount of concrete would be required (fig. 4.7). Larger sites demanded consideration of additional details. If chutes were to be used, their slope and length had to be such as to prevent the separation of materials in transit. For the same reason, wet concrete could not be allowed to drop from any great height. In the achievement of speedy, efficient reinforced-concrete construction, the handling of concrete became an engineering task as important as building design.[48]

In describing attempts to rationalize the construction site after 1900, it is important to note that the trend towards mechanized construction never superseded builders' concern for economy. For example, despite the rapid onset of mechanized construction techniques after 1900, as late as 1931 horse-drawn wagons were still being used where motor trucks would become mired in the deep mud of construction sites and waste valuable man-hours. Similarly, if a site was too uneven to allow planks to be placed directly

---

FIGURE 4.6. *(opposite)* Bucket and belt conveyors for use on concrete construction sites. Technologies for the continuous movement of sand, gravel, cement, and wet concrete around the building site imitated those employed in factory-based manufacturing processes and suggest that modernizations of work organization and management occurred in the ostensibly "craft-based" world of building. *From Smoley,* Stone, Brick, and Concrete

FIGURE 4.7. The rationalization of concrete handling and the exploitation of its fluid character inspired the creation of portable concrete-mixing towers that moved entire preparation processes around the building site. Builders created the tower and track system shown here for erection of the Grand Central Terminal in New York between 1903 and 1908. Immense systems of towers and automated craneways were instrumental in this period in the construction of the Panama Canal, portions of the New York City subway system, and similarly ambitious concrete projects. *From Halbert P. Gillette and George S. Hill,* Concrete Construction Methods and Cost *(New York: Myron C. Clark, 1908)*

on the ground, it was often uneconomical to use systems of wheelbarrow ramps, and hand-shoveling would be preferable: "The cost of constructing a runway supported by posts will often equal the cost of mixing and placing per cubic yard of concrete, and it is evident that the manner in which the mixed concrete is to be transported should be given careful consideration."[49] Even on large, well-capitalized projects, builders were flexible in their embrace of mechanization. The selective use of machines over human labor may remind us of the resistance shown by materials experts to the mechanization of test-specimen preparation. But we should be clear that any limited commitment to automation shown by the building firm managers arose not from a desire to preserve skills but rather to save costs.

Managerial Developments in Factory Construction

Organizational changes accompanied the mechanization and rational arrangement of the construction site as a "mass-production" operation. As Alfred Chandler has written of manufacturing in the period after 1900, organizational changes brought about innovations in the structure and control of the activities of workers and managers. Chandler explains that "the coordination of high-volume flow through several processes of production led to the hiring of a staff of salaried managers" who made decisions about the allocation of labor and resources, the acquisition and routing of materials, the division of labor, scheduling, and many other logistical matters in the factory.[50] Building firms that erected reinforced-concrete structures for industry between 1900 and 1930 operated along these same lines. They created formal guidelines and reporting procedures for routine activities, techniques considered to be hallmarks of modern management.[51]

The operations of the construction site were administered with what appears to have been a substantial awareness and achievement of the tenets of systematic management. For example, William Haber's 1930 report on American construction trades warns against "nonscientific" methods of project management: "No study has been made of the amount of time lost by workmen through failure in material deliveries, but from the meager evidence available it seems to be tremendous. With the same modern scientific organization in charge of construction, the contrast between its operations and those of the 'broker' contractor becomes more striking."[52] It is not simply the size of the integrated firms that brings them success, but their scientific nature; Haber's use of the word *broker* may carry an intimation of undeserved profit as well. He associates the work of small firms and independent contractors with "excessive competition" that "puts a premium on astuteness and disloyalty rather than engineering skill."[53] This disparagement of independent contractors and small building concerns was not unique to Haber. The makers of one reinforcing system advertised in 1920 that they "would not license contractors or materials dealers." They wished to place their products in the hands of acknowledged experts only. Notably, the practice of obtaining free plans for factory buildings from steel suppliers and even from insurance companies was also losing favor among factory owners. The erection of the efficient, economical factory building was coming to be associated with firms that were at once specialists in this type of construction and integrated enterprises able to handle every aspect of the task.[54]

This type of expertise, embodied in a multifaceted corporation, did more than disparage competition; it eliminated competition in many instances by its monopolistic nature. However, in explaining the popularity of the large firms it should be remembered that they could not have displaced existing approaches to factory construction had they not created a demand for their particular approach. A major tactic used by firms in this self-promotion was to distinguish the expertise of the specialized factory designer and builder from that of the building's owner. One specialist in factory design who favored the large firms suggested in 1908 that "the processes involved in all manufacturing plants and their correlation or what can be called the plant geography are so completely engineering matters that before selecting the site for a new plant an engineer should be engaged."[55] Another engineer reminded owners that when they selected established engineers to design and erect their plants, it would be clear that "the creative work of the industrial engineer has to do with such matters as are not usually included in the routine experience and work of owner or operator."[56] The founder of a large factory-engineering/building firm, promoting his company in 1919, cast the relationship between industrialist and industrial engineer as similar to that of client and attorney. The analogy suggests that the knowledge of the engineer was specialized, necessary, and of the highest professional caliber.[57]

According to this formulation, while construction could involve the same organizational methods that manufacturers used, specific skills were not necessarily transferable between different production situations. Large engineering firms also found customers for their factory-building services by distinguishing their expertise from that of other types of building firms, which involved articulating their unique abilities in an environment of standardized production. At times this claim took the relatively straightforward form of criticism of less accomplished practitioners. Clayton Mayers, an engineer for the Aberthaw Construction Company, described in detail the possible errors that could occur in beam design. He warned of excessive and inappropriate reinforcing practices, specifying that "these errors are not errors in computations, but are errors of careless design and the result is dire waste of materials." Mayers made a careful distinction between the theory and the practice of reinforced-concrete construction. In so doing, he blamed uneconomical results on the mistakes of certain practitioners rather than on the expense of hiring specialists who could, if properly qualified, repay their clients with efficient, high-quality work.[58]

Mayers's warnings continued with a second, more subtle but also more sweeping caveat. Like other engineers of the day, Mayers pointed to the need for the assistance of knowledgeable professionals in following the growing body of codes and standards for concrete. As far as city building codes and inspection systems were concerned, an owner risked great waste by relying on city authorities for economical construction: "Whoever heard of one of these authorized engineers returning a set of plans with suggestions for a more economical design? It is not the function of City Building Departments or their engineers to look for economies or suggest savings."[59] Standards, and the whole body of standardized systems of reinforcing and concrete construction, posed similar dangers to the economizing designer and owner, because as Mayers pointed out, each building presented "new problems." Only by careful study could the designer achieve effective and economical application of standardized products and procedures.

This reasoning served the professional interests of engineers and other factory design specialists. Members of these occupations commonly claimed that "materials alone do not constitute a system." A. J. Widmer, a consulting engineer who specialized in reinforced concrete, wrote in 1915, "A staff of experienced engineers is a most essential feature of a true system. The furnishing of reinforcing steel of correct types cannot constitute a system unless the design of the structure is completely in the hands of engineers experienced in the application of those particular types."[60] According to the engineers, savings were to be had from eliminating the need for skilled labor on the work site, not from eliminating consulting engineers. Advertising rhetoric combined claims for the efficiency and speed of building systems with invocations of "proved experience." As another prominent engineer put it in his 1911 prescriptions for construction using standardized elements,

> The assembling of these materials into final structures and the installation of the equipment would be under the direct control of those who know the exact reason for the provision of every single feature; and their knowledge of future operating conditions enables them to exercise an intelligent discretion that should result in a more harmonious whole than could result solely through a literal adherence to the most elaborate specifications.[61]

Once again, implementing a standard actually involves discretion and flexibility on the part of the expert who interprets it. These engineers attempted not to discredit standardized construction but to assure their own involvement in it, a strategy common to the technical professions of the period.[62]

The combination of standardized materials and methods with customized applications could have struck building owners as paradoxical. Why did building systems and standards exist if not to eliminate the necessity for (costly) specialized expertise? In the trade literature of the early 1900s, the specialized experts countered this argument with a sophisticated description of how science could work for commerce. The overview of productive work that experts in technical fields maintained was depicted as a source of continuing technical refinement, a distinct category of knowledge necessary for the creation and improvement of new productive methods. Consulting engineer Willard Case articulated in 1919 the relationship between standardized technologies and engineering expertise. He noted in the recent development of factories "a logical and healthy tendency from several causes toward *type classification,* and this has embraced not only the form of design and character of construction, but the exterior architectural treatment as well" (emphasis added).[63] To refer to engineering and construction work in this way elevated it to the status of a scientific pursuit and made standardization seem not a reductive simplification of labor but a complex analytical undertaking akin to biological taxonomy, perhaps. This definition supported the claims of factory specialists that industrial plants "are now based on a logical scientific method of analysis" and that "the business of the engineer is the science of building."[64]

Resistance from "Below"

In their solicitations of industrial clients, the new factory-building firms clearly presumed that there was little reason to preserve conventional building methods or, by extension, conventional organizations of building labor. This study does not undertake a comprehensive review of workers' reactions to the rise of concrete construction, but it is important to register the nature of the resistance that traditionally skilled laborers presented to the self-fashioning efforts of modern engineering firms. With the privilege of hindsight we know that concrete did "win" as the favored medium for large-scale construction in the United States, but that skilled labor groups in the building trades also retained a measure of economic power throughout the twentieth century. We can suggest here some features of the dynamic relationship engendered by the use in many instances of concrete instead of wood, brick, or stone.

By the middle of the nineteenth century much of the labor of construction was done by workers organized into guilds, unions, and less formal family and community networks. Even the weakest of such carpentry, masonry, or metalworking organizations could to some degree control local wage levels, work pacing, and availability of skilled labor. In larger cities the power of trade groups was sometimes linked to that of political machines and almost hegemonic. When large concrete firms after 1900 offered the integrated services of factory siting, design, and erection, they helped building buyers sidestep involvement with many such labor groups. Here they followed a precedent set by large firms of the late nineteenth century that brought their own brickmakers, stonemasons, and carpenters to different locales. The adversarial features of this centralization are obvious; the head of one bricklayers' union referred to such corporate conduct as an "invasion" of a locality.[65]

The resistance offered by trade groups to such strategies was many-sided. First (and clearly ineffectively, in the long run), representatives of carpentry and bricklaying groups tried to discredit concrete as a safe and economical material. In the journal of the United Brotherhood of Carpenters and Joiners of America, editors reported a "Public Getting Wise to Wood Substitutes" and recorded the "complaints" of a waitress lately employed in a concrete-floored restaurant: "I don't know what is the matter with me lately. I have done this work all my life but since I came to work to this place I am so tired at night I can hardly move."[66] Addressing matters of skill, advertisements and articles for wood materials and brick claimed that such technologies, unlike concrete, used "talented" and "out of the ordinary men." One essay published by the Building Brick Association of America seemed to be answering the scientific rhetoric of the cement industry quite directly: "There is nothing new-fangled about brick—it is not an experiment and has no unsolved mysteries—brick is on every hand and thoroughly understood by the builders in every community." Pointedly, the Bricklayers and Masons International Union (B&MIU) vowed in 1904 to report in its journal every incidence its members encountered of "cracking, breaking or total collapse of concrete" and credited the "best architects and engineers" with comprehending concrete's inferiority to brick, stone, and terra cotta.[67]

In large measure such rhetoric was aimed at people already ill disposed towards concrete. More effective were the attempts of bricklayers, stonemasons, and practitioners of other established crafts to hinder the operations of the concrete industry by refusing to work on projects deemed unfriendly to union interests. "Stay-away" orders were common in the first decades of the century among all types of trades seeking to punish non-union employers, and they were a potentially useful device in construction in times of economic health (in depressed periods union members were understandably less likely to turn down work on such a basis). Most buildings used a combination of materials, and members of the Bricklayers International announced more than once that their brotherhood would refuse to lay bricks on any building in which concrete was also employed. Because brick buildings by the 1890s commonly used cement and concrete for fireproofing and foundations, the bricklayers could have had a major impact with this stay-away scheme had enough individuals participated in the effort. However, union leadership at the highest levels dictated a different response to the building industry. In 1904 the American Federation of Labor (AFL) granted a charter to the American Brotherhood of Cement Workers, recognizing either the distinct physical or distinct economic nature of the emerging technology. Without the support of the powerful parent labor organization, bricklayers could hardly hope to establish control of concrete work. By 1906 the B&MIU was already conceding that members would accept work on concrete projects if granted supervisory status for any concrete use on the building site. This gained the bricklayers some ground. The National Fireproofing Company, among the large employers most friendly to the bricklayers' union, began using the Kahn System of concrete floor arches rather than brick infill at this time, and the union won the right to "remain and supervise the placing of concrete."[68]

But as already discussed, the potential of concrete systems to cut wage costs was great, and the bricklayers' chances of retaining such an anomalous role in concrete construction were slim. In 1923 the AFL turned control of "artificial stone"—meaning precast concrete elements—over to building firms, definitively distinguishing concrete from traditional masonry methods, and any hope of skilled craft control over the medium was largely lost with this concession to employers' economizing impulses. Significantly, the dispute was not one divided strictly along management/labor lines. The building industry has always been extremely sensitive to economic fluctuations, and in these disputes the jurisdictional stakes were clear: Bricklayers, stonecutters, and stonemasons had fought among themselves for decades over which craft should handle each new type of tile or cladding. With new technologies determining employment opportunities, the difference of three-quarters of an inch in width might define a tile as belonging to the category of "brick" rather than "stone" and form the basis of extensive debate. But cement and concrete carried a new set of implications for these groups. As a medium predicated on a reduction of human labor, concrete foretold not a redistribution of skilled employment opportunities among the trades but rather their elimination. The president of the B&MIU had almost predicted this outcome in 1905: "When our unions gave away the control of the installation of concrete fireproofing, they made one of the costliest and most serious mistakes in the history of the craft."[69] Concrete's historical identity here is not simply one of "technological revolution" but one of shifting economic dominion between labor and employers in the building industry.

The Aberthaw Construction Company: A Case Study

A study of the policies and practices of the Aberthaw Construction Company, a successful engineering/building firm that specialized in reinforced-concrete factory construction, will illustrate how such firms translated a knowledge of new construction techniques and the application of scientific management into commercial success and, more important, an increase in status for themselves and the concrete factories they built. Aberthaw was a functionally integrated engineering/building firm.[70] On some occasions the company would work with other design or engineering firms, such as Lockwood, Greene, but for the most part it initiated and completed factory construction projects by employing solely its own staff. In many respects Aberthaw's practices were typical of large firms of its day in relying on both standardized building products and procedures and elaborate bureaucratic structures. Some of the company's directors had had university training in the methods of scientific management, and on at least one project Aberthaw employed consultant Sanford Thompson, a colleague of Frederick Taylor who specialized in time studies.[71] The company's history illustrates the interaction of technical choices, management styles, and labor relations in factory construction between 1900 and 1930.

The Aberthaw Construction Company was founded in 1894 in Maine expressly to specialize in concrete construction. By 1902 the company had relocated its headquarters to Boston and received a commission to build the Harvard University stadium, the first reinforced-concrete stadium built in the United States. Other early projects proudly advertised by the company included the first concrete sidewalks in Boston to incorporate glass skylights (for cellar illumination) and reinforced-concrete buildings for the Navy Yards in Charleston, Massachusetts.[72] Aberthaw's purchase of regional patent rights for Ransome's steel reinforcement designs positioned the company well in the growing market for utilitarian concrete buildings.

Although the company occasionally accepted a commission for a residence, its primary interest was in industrial structures. These included not only factory buildings but accompanying retaining walls, coal pockets, storage tanks, chimneys, and other structures for which reinforced concrete was well suited. Like other firms that made a specialty of building factories, Aberthaw advertised its ability to accommodate a variety of clients. A 1915 catalog shows completed buildings for the Carter Ink Company (Cambridge, Mass.), the Pierce-Arrow Motor Car Company (Buffalo), the Bridgeport (Conn.) Brass Company, Pacific Print Works (Lawrence, Mass.), and the Hood Rubber Company (Watertown, Mass.), as well as factories for firms that manufactured lamps, toothbrushes, wire, creosote, stationery, and a variety of other products.[73]

By 1920 Aberthaw had major offices in Boston, Atlanta, Philadelphia, and Buffalo. A striking feature of Aberthaw's catalogs of the pre-1930 period is the uniformity of the product the firm offered. The many different manufacturing facilities Aberthaw built bore a remarkably similar profile. Most were three to seven stories tall and displayed an exposed concrete skeleton.

Ornamentation was virtually absent. This austerity suggests the methodical mass-produced character of Aberthaw's buildings. At the same time, the company used advanced scientific practices of the period constantly to adjust its technological procedure for greatest safety and economy. For example, as early as 1902 the firm's engineers submitted reinforced-concrete beams for testing at the Massachusetts Institute of Technology. Brochures included photographs of dramatic tests performed for Aberthaw in buildings under construction.

"No Lost Motion, No Waste of Time"

Aberthaw's managers accomplished the erection of uniform, high-quality buildings with advanced organizational methods. As Aberthaw described itself in 1918, the company was the epitome of systematic management: "The organization is the successful co-ordination of many elements, both human and material . . . not a mere aggregation of individuals. . . . Management and operation is divided in such a way that there is perfect coordination. . . . [There] is no lost motion,—no waste of time or effort."[74] This administrative coordination was achieved by maintaining a central office from which instructions for all projects emanated. An elaborate hierarchy of managers and workers allowed Aberthaw to take on immense jobs, one reaching a cost of $20 million and requiring a force of fifteen thousand men. Each project was assigned a superintendent, below whom worked a chief engineer and a chief clerk. The chief engineer established schedules and supervised in turn a scheduling engineer, a purchasing agent, and a head of an employment bureau for the project. Most projects employed a separate field engineer, who would supervise carpentry, steel, mortar, and concrete foremen and a master mechanic, as well as assistant field engineers functioning as inspectors. The chief clerk monitored costs. Once the project was under way, the job superintendent would assign gang bosses to supervise workers directly.[75]

The functional departmentalization of work at Aberthaw echoes the standard corporate management techniques of the day.[76] The company's methods for keeping track of what work was being done when, and by whom, were also corporate in nature. Since the mid-1880s, manufacturing firms had been refining systems of "shop-order accounts" in which orders were numbered and assigned routing slips, and all materials used and operations performed in filling the order recorded on those slips. By 1920, texts on the management of manufacturing and construction businesses explicitly referred to time schedules, working estimates, and daily reports and diaries as control instruments.[77] In Aberthaw's case it was tasks on the construction site, rather than product orders, that became subject to such scrutiny. Because the word *field* sometimes precedes an employee's title in Aberthaw's project records—as in *field superintendent*—and sometimes does not, it is difficult to ascertain which employees worked in the company's central offices, which in the field, and which moved between the two locations. However, the system of detailed time and cost sheets maintained for projects after about 1914 shows that there was certainly a flow of information from the field to administrative departments.[78]

Edward H. Temple, who began working for Aberthaw in 1911 and eventually became the firm's general manager, inaugurated a number of these procedures. On the Pacific Print Works project of 1912, Temple decided that he wanted to keep track of running costs "just as a doctor would keep a fever chart."[79] Unit costs for individual tasks, such as placing column reinforcement or pouring floor slabs, were recorded as the job proceeded. The resulting records, called "bogeys," gave the company a ready basis for monitoring expenses. Costs could also be projected to completion. Eventually Temple persuaded the company to use an alphabetical code to track work undertaken. He used a capital letter to indicate the general type of work (*B* for brick, *D* for digging, *M* for concrete, *F* for forms, etc.); a lower-case vowel for the specific task (*e* for assembling, *i* for stripping, *o* for repairing, *u* for unloading); and a final consonant for location (*f* for floor, *r* for roof, *w* for walls): "From the above explanation it can readily be seen that *Mef* always means placing of concrete on the floor; *Fef* erecting forms for floors; *Bew* erecting brick walls." According to Temple, new employees quickly became fluent in the use of these codes, which appear in some but not all surviving work records.[80] The system seems fairly absurd in its complexity, but it reflects Temple's faith in the ability of such translations to regularize both communication and labor.

An example of Aberthaw's administrative sophistication is the firm's erection of an arms plant during World War I under tremendous time pressures. The government called on Aberthaw to build a new plant for Colt's Arms in Hartford, Connecticut, and the firm managed to complete the job in just forty days. For this project, elaborate schedules were drawn up for sixty-seven different items, such as clearing the site, general excavation, casting concrete floors and columns, and casting sills and coping. Three copies of each schedule were sent to the job: one for the superintendent, one for the routing department, and a third to be kept up to date weekly and sent back to the Boston office. A fourth master copy was posted on a bulletin board in the Boston office to record all progress. Despite the rush, Aberthaw described its methods at the Colt's plant as standard for the firm: "It was a case of economy in time due to systematic handling rather than to any particular speeding up."[81] The company thus sought to dispel any concern that its speedy erection of factories involved haste or negligence.

The Worker in the System

While Aberthaw's managerial staff kept close track of expenditures of labor and money, they also concentrated on refining the routing of materials on the construction site. This was a crucial undertaking for a number of reasons. Although other records described the movements of workers, the control of materials functioned to constrain those movements. For example, if supervisors wished a certain amount of reinforcement to be put in place on a given afternoon, they could provide that quantity to a location and have a crew remain at that location until all the reinforcement had been used. Or supervisors could check at the end of the day to see what portion of the allotted materials had not been placed, identifying insufficient activity by workers. If work orders indicated the intended movements of workers, and record cards their supposed accomplishments, the availability and consumption of materials on the site actually determined and revealed their movements, respectively. The quantity and location of materials on the building site provided both a control over and an index of work.

By 1900 the routing of materials was becoming a major concern for many manufacturers. Treatises on systematic management always included advice on the layout of factories to accommodate the delivery, storage, handling, and shipping of raw materials and finished goods.[82] But the construction industry experienced some difficulty in applying this advice to building sites. Building materials were often very bulky, weather could prohibit storage on an outdoor site, and urban locations frequently offered limited free space on a site. Aberthaw attended very carefully, therefore, to the role of materials in planning a construction project. Each step in a building's erection was mapped out in terms of the acquisition and delivery of materials and their distribution around the site. Very early in any job the superintendent completed blank forms for every material or piece of equipment required. Using a master "Progress Schedule," office staff ordered materials to arrive on the site just before the date they were needed, and a yard staff monitored all stock. The company's 1931 instruction booklet indicated the importance of materials flow for the timely and effective execution of virtually every construction task.[83]

As general manager Edward Temple put it, "All this is much like a railroad time table showing when the train must be at intermediate stations to pick up people coming from the side lines."[84] Aberthaw's emphasis on routing had a strong impact on the nature of work on the construction site. First, Aberthaw's managers claimed that "good records [are] the best possible basis for advancement and permanent employment." In essence, the bogeys and other evidence of materials consumed eliminated subjective judgments about worker performance.[85] Second, in linking measures of productivity to materials consumption, this system treated the person handling the materials as a consumer of materials only, rather than as a working person with mind and body. In both respects, considerations of the physical difficulty of the work, the degree of training or experience required, and working conditions disappear from the analysis of how labor should be managed and compensated. (At least one Aberthaw brochure happily informed prospective clients that "weather was no object" on its jobs.)[86]

Aberthaw's managers did not display a total disregard for their workers' well-being, and additional features of their "worker-welfare" programs will be explored in chapter 5. The company had for some years a gain-sharing plan that correlated increased worker output with worker rewards. Gain-sharing had been used by industries since the mid-1880s to provide an incentive to workers.[87] It involved returning to workers some portion of any extra profit that derived from their accomplishment of work under budget or ahead of schedule. Aberthaw began its bonus system on its Pacific Print Works project of the early 1920s by offering workers and their foremen 50 percent of any savings on the predicted costs. Soon this system was extended upward to include higher-level employees such as superintendents, chief clerks, and engineers. Aberthaw believed the system to be successful in part because it prompted workmen to push "everybody concerned to get the materials to them—not when wanted but a little ahead of time." In this way Aberthaw derived economies from many employees while giving bonuses only to some, at the same time counting gain-sharing among the favors it provided workers.[88]

Moreover, as labor groups and analysts of the subject have established, monetary incentives do not equate with worker authority.[89] That Aberthaw believed employer-employee cooperation of this sort to be a means of controlling rather than empowering workers is suggested by the other ways in which the company tried to limit workers' self-determination. Until 1929 Aberthaw prohibited union involvement on all projects and divisions. The company's managers claimed that in the absence of union regulations, they could recognize individual merit, pay only what a worker's ability indicated, and adjust hours to the requirements of a project. They claimed also that because they did not have to comply with union demands, their task and bonus system truly provided an incentive for extra effort and lowered unit costs.[90] Aberthaw also argued that its policies served the public good. During World War I, Aberthaw erected the Squantum Destroyer plant near Boston with nonunion labor and claimed to have thus saved vital defense dollars.

In the late 1920s, however, the company found it difficult to hire the needed workforce when it undertook projects near large cities with entrenched trade organizations, and in 1929 it began admitting union workers to its operations. The company's distrust of the unions' attempt to control work processes is demonstrated by a 1931 instruction manual that specifies work procedures: "Swivel power charging hoppers are useful in cases where a concrete mixer cannot be charged[;] . . . the use of this type of machine necessitates (according to union rules) an additional hoisting engine."[91] The parenthetical addition suggests that but for union rules, no additional hoisting engine would be needed. The union's requirements appear to have superseded Aberthaw managers' own judgments about where and how to apply machinery and undermined the meticulous workplace control they had previously achieved and under which construction technologies had taken shape. The period during which rationalized construction techniques automatically reinforced centralized management of labor was over.

Creating a Public Image

While Aberthaw's directors of 1900 to 1930 sought to establish controls within the workplace, they also faced pressures from the marketplace. Since their ability to produce high-quality buildings could not by itself guarantee Aberthaw's success in a competitive environment, they addressed the problem of establishing a favorable reputation for the company. Moving beyond simple advertising of its services, the firm turned to progressive business methods—including systems of charging for its services—that distinguished its work from that of other firms.

Aberthaw's directors were apparently aware of the latest theories of business operation, which had identified several dangers in the competitive practices by which contractors and building firms procured work. Until about 1920 the most common type of contract between owner and builder was the lump-sum contract in which the builder figured a set cost based on plans and specifications and charged the owner accordingly. This system had drawbacks for both builders and clients. To protect themselves from unforeseen problems in the course of construction, builders often built extra charges into their lump sums. Further, in seeking a builder, owners could solicit as many bids as they liked. This meant that suppliers would seek to undercut one another and might lower their bids to the point where they could not hope to turn a profit without somehow shortchanging their clients. This might be through hidden scrimping (say, "the contractor 'forgets,' under such competition, to paint the beams according to the specifications"). Or it might entail overcharging on another job. In general, unchecked competition was believed by many in the industry to drive standards of work and craftsmanship downward.[92]

William Haber, in analyzing the building industry in 1930, described three possible solutions to this problem. First, suppliers of materials and services could agree among themselves to charge for estimates and thus discourage buyers from "shopping" bids and driving prices down. Second, buyers could solicit a "quantity survey," giving a single independent estimate of materials and costs for use by all bidders. In this way bids could be easily compared, and artificially low ones could be eliminated from the competition. A third possibility was to institute a "cost-plus-fee system," and this was the method used by Aberthaw from the 1910s onward. The cost-plus-fee, or percentage, system called for the owner to pay all costs incurred in a construction project, plus a fee or percentage of those costs. Bidding was eliminated. As Haber put it, this method put an emphasis on "service." Suppliers did not scrimp, and the owner was fully aware of the builder's profit margin.[93]

Some observers believed that the cost-plus-percentage system led builders to run up costs knowing that they stood to profit from every dollar spent. For this reason, some firms chose to work only on the cost-plus-fee system and argued that the choice reflected the high quality of their work. One reinforced-concrete company claimed in 1907,

> While the advantages of good concrete construction are obvious, the dangers resulting from poor work are very great. A little slighting in the quality of the cement, a little skimping in the richness or thoroughness of the mixing, or a small percentage shaved from the weight of reinforcement change the entire result of the work. It is easy for a contractor, when he sees his profits vanishing, to do a little skimping here and there. . . . The insidiousness of this temptation is the reason why we seek work only on a cost-plus-fixed-sum basis.[94]

Aberthaw followed this same tack, declaring that any bidder could, by studying specifications very carefully, detect ways to substitute less expensive materials or methods for those specified. The firm performed 90 percent of its work on a cost-plus-fee basis and believed that this policy proved its integrity.[95]

According to Aberthaw's directors, using the cost-plus-fee system enhanced the firm's reputation because the system created an atmosphere of "mutual confidence" between builders and owners.[96] Many engineers and building firms believed that under the cost-plus-fee system, client relations were elevated from a merely commercial plane to a professional one. Edward Temple asked why people accepted that doctors, lawyers, and dentists worked on a cost basis but expected "to get good building work done, which is also pure service, on a low competitive basis." He wished factory construction to achieve the status of other respected technical enterprises.[97] Aberthaw's promotional materials echo this goal. They describe the advanced production techniques of the manufacturing companies for which Aberthaw built factories and characterize Aberthaw's own methods as the only means of achieving appropriate physical settings for these forward-thinking enterprises.[98]

Together, such "progressive" advertising rhetoric, cost-plus-fee charging, and systematic organization conveyed to clients Aberthaw's participation in modern business management. But the most obvious advertisements of Aberthaw's modernity were the buildings it erected. The generic reinforced-concrete factory building, bearing no ornamentation or other reference to conventional architectural practice, proclaimed itself a new invention made by new methods. The appearance of such a structure declared its builder's and owner's allegiance to the processes of modern commerce—including the stratified, rationalized building processes that were replacing conventional artisanal building skills—and celebrated these processes as a basis for architectural style. Simply put, the reinforced-concrete factory building represented a new concept of good taste. When Aberthaw and other builders commissioned luscious full-color paintings of factory buildings to reproduce on their brochure covers (frontispiece), they were implying that the factory building deserved respect as a cultural icon. The following chapter will ask if factory buildings received such respect, and what social and cultural values the reinforced-concrete industrial buildings may have represented for their owners and critics.

Conclusion

In his 1959 comparison of bureaucratic and craft administrative methods, Arthur Stinchcombe defined mass production as production in which the following determinations were made by managers rather than workers: the location of work; the movement of tools, materials, and workers; the particular movements of a task; schedules and time allotments; and inspection criteria.[99] Managers of the Aberthaw Construction Company and other firms like it dictated these aspects of work to their workers, systematizing technologies and administrative methods in order to establish reinforced-concrete construction as a mass-production process. The reinforced-concrete factory building of 1900 to 1930 was the standardized product of a highly rationalized segment of the construction industry.

In following the tenets of standardized production, the engineers and builders who specialized in reinforced-concrete factory buildings were careful to preserve their own privileged status. As was the case with materials testing and regulation for concrete, systematized construction threatened to undermine its promoters' own claims of technological and managerial authority. Thus, even as engineers developed and deployed systems of prefabricated reinforcement and precast concrete that simplified and deskilled the labor of construction workers, they raised the flag of their own technical expertise. They argued that their knowledge of classes and types of buildings gave them a superior understanding of clients' needs, and they sought to imbue the erection of concrete buildings with a scientific character. At the same time, the rationalization of the construction industry undermined the power of (nonunionized) workers and made it easier for managers to limit workers' comfort and mobility for the greater profit of their employers or their employers' customers.

In its treatment of workers and other administrative choices, Aberthaw embraced contemporary business methods usually thought by historians to be more typical of manufacturing firms than of construction businesses. Creating a public identity for itself as a thoroughly modern operation, Aberthaw achieved success in the factory construction market. Compare the appeals made by the forward-looking concrete construction company with that of a concern that produced tin plate in 1906: the American Sheet and Tin Plate Company claimed that its process "is the oldest of the 'old style' methods and [our] plates are made today just as they were nearly a century ago."[100] As proponents of brick had also claimed, convention represented proven knowledge. Then as today, tradition and modernity variably served to construct appropriate public identities for different products. In keeping with its forward-looking stance, from 1900 onward Aberthaw's buildings showed little deference to architectural traditions and celebrated all that was new about their form and their means of construction. That newness included all the emerging methods for rationalizing the processes of construction. Thus, in their frank expression of the conditions of mass production, reinforced-concrete factories proposed new architectural values based on the modern social relations of a capitalist nation. To comprehend the popularity of the unadorned, standardized reinforced-concrete factory building in the United States after 1900, we must look at the relevance of that modern social ideology to those who commissioned these buildings. This inquiry forms part of a larger one: How did a series of technological and industrial agendas come to shape the American architectural canon? The social and cultural character of the reinforced-concrete factory building is the subject of chapter 5.

CHAPTER FIVE

What "Modern" Meant
Reinforced Concrete and the Social History of Functionalist Design

> If an engineer, meeting a special problem in a purely scientific way, produces a building of beauty, he has produced architecture. . . . He becomes—temporarily, at least—an architect.
> —G. H. Edgell, 1928

For all the ways in which reinforced-concrete buildings typify modernizing industrial practices after 1900, they bear an important distinction from assembly line products and mass-produced machinery. These are buildings and therefore have an identifiable aesthetic character. All human-made objects may be said to have a "design," and it is arguable that even the most mundane product has been intentionally placed by its makers on a time line of expressive idioms. But concrete buildings are "architecture" and are therefore readily placed among a set of expressive conventions with particularly strong cultural resonances. The vast majority of reinforced-concrete industrial buildings between 1900 and 1930 were apparently lacking in any architectural ambition—austere, standardized, unremarkable by any conventional measure of aesthetic accomplishment. But bringing to bear on these structures the complex social origins and consequences described in previous chapters, I here argue that the supposed absence of aesthetic intention seen in the standardized concrete buildings actually represents a set of deliberate and constructive choices on the part of their designers and buyers. The functionalist form given to concrete buildings elevated new technical and social practices to the level of high-culture accomplishment.

Cultural influence arises from and in turn lends social power. The acceptance of standardized, functionalist building forms among factory designers and owners had its origins in the hierarchical social program we have already identified. The uniformity of concrete factory buildings celebrated the comprehensive, "sciencelike" thinking that many professionals believed grounded other tasks of standardization (such as the creation of technical specifications). Further visions of industry as a progressive moral force in modern society fueled this commitment to modern architectural design. As functionalist concrete buildings began to dominate industrial landscapes after 1910 or so, their credibility as cultural forms expanded. The consequences of aesthetic influence can reach far beyond the narrow, sometimes self-referential discussions of architects and critics to confer social authority.

As they pursued modern building methods, the factory builders of this period pursued a modern appearance for their structures. Turn-of-the-century builders of office buildings and showrooms retained the eclectic decorative appearance that prominent commercial structures had shown throughout previous decades, even as they used new methods of steel and reinforced-concrete construction. But factory builders—by this I mean building firm owners and operators, and consultants in the field—turned to a new appearance for their buildings as readily as they turned to new materials and means of labor organization. The majority of reinforced-concrete factory buildings erected after 1900 bared their gray concrete framing to the world and offered no cladding, ornamentation, or other distraction from their modern structural character. As shown in earlier chapters, many of the factories were designed by engineers working for building firms, without the participation of academically trained architects. These factories bluntly expressed their origins in the streamlined and standardized procedures of engineering.[1]

Study of builders' attitudes towards reinforced-concrete factory buildings reveals a positive rationale for the "negative" phenomenon of architectural simplification and standardization. The structures represented to their builders a fiscally sound renunciation of historicist ornamentation and conventional artisanal building skills, but not an absence of aesthetic doctrine. The plainness and overt standardization of the concrete structures reflect instead a willingness to publicize the achievements of mass production, which included the activities occurring inside the factories and the materials and techniques of reinforced-concrete construction. The stark structures express the belief that the workings of industry need not disguise or embellish themselves to gain public approval. Their builders maintained that the reinforced-concrete factory buildings constituted a redefinition of architectural accomplishment: an elevation of production and commerce to a level of cultural prestige previously occupied by historicist architectural motifs.[2]

The modern factory building emerged not only from a spirit of artistic reform but from a type of social progressivism as well. In their eagerness to install the methods and benefits of mass production, factory builders came to believe that the modernization of the industrial workplace necessitated new consideration of workers' experiences. Many factory-building firms, extending their services beyond the strictly technological, offered their clients advice in the area of worker hiring and administration, services that in their pursuit of an orderly and hierarchical workplace echoed the building firms' own internal management methods.

To describe the events that accompanied the erection of reinforced-concrete factory buildings, I begin by tracing the builders' understanding of contemporary critical frameworks in architecture and their two-sided conception of the factory's place in the architectural canon. Standing outside the sphere of academic architectural practice, working with a new set of materials and methods, the builders sought with their designs to fulfill conventional aesthetic criteria of beauty and harmony. At the same time, they sought sanction for a distinctly new architectural type. By their builders' own descriptions, reinforced-concrete factory buildings were intended to look in most respects unlike other structures, to derive their form from their function as modern commercial venues. A segment of the academic architectural world offered support for this program, and the two groups together created a congenial public identity for the new structures as additions to the architectural canon.

I next set the factory builders' conception of aesthetic modernity into a framework of social change. Large factory-building firms embraced contemporary managerial trends that saw the safety and comfort of workers as a responsibility of employers. The builders made factory workers' health and contentment part of their "product line," offering physical plants that assured buyers of both. This paternalistic approach reached beyond the factory building itself; employee housing became another area in which the building firms offered their services. In this area some of the complexities of modern industrial employer/employee relations come to light; the social and moral uplift of workers promised by factory builders carried intimations of social control as well. The work of the modern factory-building firms embodied a mixture of social effects.

This chapter concludes by posing a question for future study: What might the functionalist reinforced-concrete factory buildings have represented aesthetically to the industrialists who commissioned and occupied them? In commissioning standardized reinforced-concrete factories, thousands of American manufacturers associated their firms and themselves with functionalist design. By 1930, buyers of reinforced-concrete factory buildings had subsidized a transformation of large parts of the American industrial landscape into an environment of stark and uniform structures that celebrated the standardized cultural product. What understanding did factory owners of 1900 have of these architectural transformations? How broadly did they conceive of the symbolic import of functionalism?

In tracing the outlines of the aesthetic and cultural discourse that surrounded reinforced-concrete factories between 1900 and 1930, we can begin to see some of the reasons for functionalism's enduring popularity in twentieth-century American building design. Many mid- and late-twentieth-century architects have drawn inspiration from the functionalist buildings erected in the United States after 1920 by European interpreters of the style. The "International Style" architects themselves are known to have admired American concrete factory buildings.[3] But the American enthusiasm for austere, highly standardized buildings predated this academic practice. For many builders and building buyers, the aesthetic acceptability of standardized structures may have had its foundation in the ranks of solid, imposing concrete factory buildings that spread through the country after 1900.

The Standardized Factory Building and Contemporary Architectural Tastes

The reinforced-concrete factory buildings that began to appear in American cities and suburbs after 1900 did not look like existing commercial or civic buildings. They displayed none of the ornamented, eclectic styling that had predominated in heavily capitalized American structures throughout the last quarter of the nineteenth century and that still characterized many new factory buildings. Even the color and surface texture that brick cladding could have brought were missing from most of the concrete-frame buildings (fig. 5.1). However, the building firms and consultants who designed and erected these factories were neither ignorant of contemporary architectural fashion nor dismissive of its demands. They crafted their aesthetic arguments for functionalist design against a backdrop of vigorous critical debate in the architectural press. The specific terms of ideological exchanges among critics and architects of 1900 ranged from the formalist to the moral.

FIGURE 5.1. Standard Oil Company plant, Albany, N.Y., near completion ca. 1922. *From Aberthaw Construction Company catalog, 1926, ACC Archives*

Advocates of modern, utilitarian design and promoters of eclectic, historicizing architecture accused one another of aesthetic ineptitude, antisocial behavior, and even antidemocratic intent. The sweeping nature of their concerns grew from their conceptions of how industrialization would transform American life and culture. Critics, public figures, and professionals of all kinds assessed American prospects in the new century. They rooted the progress or the imminent demise of American culture in the growth of mass production and mass consumption and the encroaching subordination of all other endeavors to these goals. Depending on the interpreter, American arts and letters—including architecture—and the pursuit of an orderly modern society could be expected to flounder or flourish in tandem amidst these changes.[4]

Builders of reinforced-concrete factories entered the critical fray to praise the austere structures to critics and to the larger audience of potential factory buyers. Although promotional literature produced by factory-building firms never failed to mention the efficiency and economy of concrete construction, it also offered explanations, praise, and justification for the appearance of the factories in answer to prevailing critical debates. The factory builders joined those analysts who claimed a favorable prognosis for American culture in the new era of mass production. Their buildings would be part of modern culture and challenge the rear-guard assumption that only conventional academic practice could yield buildings of architectural significance. The factory builders' arguments addressed all the sweeping complexities in which the architectural experts trafficked, listing advantages to the modern factory that included the "intrinsic value" of a well-designed building; the "good advertising value" of an attractive plant; and the general benefits of health and contentment for factory workers.[5] Each advantage indicates a portion of the factory builders' ideology of modernity.

Creating the case for the visual "pleasure" that a well-designed factory might bring to "the discerning,"[6] factory builders offered discussions of design in journals of the cement trade and factory management as well as in books on these technical topics. The prescriptive content of this literature might be described as an association of the reinforced-concrete factory's constitutive elements—the exposed concrete column, the standardized steel-sash window, and all the other simplified, repetitive forms typically used in this type of construction—with traditional architectural values of visual beauty and harmony. As chapter 4 made clear, the economies of concrete construction derived from simplicity and duplication of forms, and that practical correlation formed the basis of the aesthetic schemes proposed by many advocates of concrete. W. Fred Dolke Jr., of Lockwood, Greene, the immensely successful industrial construction and management firm, wrote in 1917,

> The essential and fundamental characteristic of a factory building will always be utility, but side by side with it now stands that other requisite, attractiveness. Attractiveness, or beauty, means simply good taste. There are many essentials in the construction of every building, such as windows, piers and copings, which can be so grouped and spaced, so molded, that beauty results with little extra cost. Beauty does not mean lavishment. It means simplicity and good taste in disposition of members, and in use of materials.[7]

Other engineers specified that economy demanded the elimination of projecting members, such as cornices and belt courses. They required that these traditional architectural details be replaced in factory or warehouse design by a sensitive juxtaposition of openings to masses.[8] With few exceptions, factory designers advocated a visual effect that followed from the simplest handling of concrete possible.

Similar arguments appear in the writings of contemporary architects and critics who offered the first widespread advocacy of the modern "truth-to-materials" doctrine.[9] In 1907 the *American Architect* solicited comments from architects and architectural writers on "the proper artistic expression of a concrete building unveneered in any way."[10] The fact that the journal categorized buildings in this way indicates that there were already enough such structures to suggest an aesthetic question. Most of the respondents agreed that design for the new structural method was "a problem to be solved." Perhaps because many reinforced-concrete buildings in America were being erected without the involvement of architects, the journal and its architect readers might have relished an opportunity to judge them.

The solution to the problem of concrete-building design as seen by many of the architects surveyed was an "absolute truthfulness of expression" and a complete rejection of the use of concrete in imitation of other materials, particularly stone. As one respondent put it, "Slender mullions and fine arrises in concrete are absurdities. The effects of the chisel upon the carved stone cannot be repeated. . . . In short, the whole category of lithic forms by their character and suggestion belie the process of their production in concrete, and the building so shaped and decorated becomes a contradiction of itself from top to bottom."[11] This echoes the sentiment of another architect, writing in the *Architectural Record* about the same time, that concrete, because it functioned as a monolithic skeleton frame, should never be used with curtain walls and other such "fictional expressions."[12] This fidelity to the physical nature of concrete was advocated as a kind of "realism" that also encompassed austerity, a corrective to Beaux Arts embellishment. Architectural critic Russell Sturgis wrote in 1904 of the "wholesome architectural influence" that new, utilitarian factories and warehouses offered American design.[13]

Whether or not they actually knew of the architects' and critics' endorsement of their "absolute truthfulness to materials," the builders of unadorned factories believed that architectural quality was something that could profitably be redefined for the new era of industrial production. One 1935 text on factory design, written by three "professors of industry" at the Wharton School of the University of Pennsylvania, claimed to summarize the lessons learned by builders during the first part of the twentieth century: "A building that merely serves its utilitarian purpose may be an ugly blotch. To avoid this situation, however, it is not necessary to erect extremely ornate offices and factories in which the concessions to 'art' are all too clearly evidence of 'conspicuous waste.' There is no reason why utility and the canons of good taste, balance, and proportion cannot be combined in the design of a building."[14] Possibly the authors had in mind Thorstein Veblen's use of the phrase "conspicuous waste" in his 1899 *Theory of the Leisure Class*. Veblen preferred the backs of many buildings—"left untouched by the hands of the artist"—to their ornamented facades, and historian Peter Conn has noted the influence of this text on contemporary architects and critics.[15] For these specialists in factory design, "art" was something not simply distinct from utility but antithetical to it. "Good taste," on the other hand, not only was still possible in the commercial context but was redefined by modern factory builders and proponents to include the frank expression of utility.

Factory builders and architectural critics were formulating new ideas of what constituted good design and, more broadly, what constituted contributions to American "taste" or culture. They extended an old aesthetic premise—that certain kinds of architectural forms were appropriate for buildings of certain functions—to a contemporary situation. For architects and critics, this aspect of "realism" was another matter of taste. A commentator writing in the *American Architect* in 1909 explained bluntly that "a free use of intricate detail or expensive materials in a soap factory would be mere affectation."[16] In 1921 another critic wrote more calmly that "a building should indicate by its exterior treatment and design something of the purposes for which it is intended. The indiscriminate use of decoration and color should be avoided in the design of industrial buildings."[17] Factory builders saw a larger reason for expressing through a building's form "the purposes for which it is intended." Both builders and critics believed that the material nature of a building can have as full an expressive meaning as any other architectural convention, but the builders also believed in the "advertising value of a handsome plant in the path of national travel." That value stemmed from the factory's identification with the industrial processes it contained. If the appearance of the factory conveyed economical and repetitious production methods, unencumbered by superfluous detail or disguise, anyone encountering the structure could see in it the modern attitudes of the building's operators, and thus of the nature of the work conducted within. Such buildings would have a "definite effect for good . . . upon customers and as an advertisement to those who pass it."[18]

We should also connect the appreciation that factory builders evinced for standardized forms to the more general elevation of the task of standardization as mental work. We have seen that engineering instructors promoted the creation of specifications as a challenging intellectual task, suitable for highly trained and highly paid personnel. Similarly, engineering firms that specialized in factory design claimed the erection of standardized factory designs as almost a taxonomic undertaking, not a reduction of detail but a selection and distillation. In the same vein, factory designers in architectural or engineering firms who promoted standardized building forms could pose the external uniformity of factories as a celebration of modern intellectual prowess: the ability to create a "perfect" form, a *type*. Older styles of intellectual accomplishment, in architecture and other fine arts, were predicated on an accretion of traditional cultural forms. In a culture that celebrated production, standardization carried the same suggestions of rigorous, synthetic thought.

In summary, the modern factory represented architecture in which the aesthetic and the commercial happily conjoined, each inspiring the other. The compatibility that factory builders saw between art and industry suggests a permeability to their definitions of culture and commerce; the practice of erecting modern factories seemed to constitute both at once. The openness of the factory builders' multivalent conception of architecture allowed them to see other functions for their work as well. As they lauded the aesthetic quality and advertising value of the modern factory building, factory builders added as a third element to their discourse on architectural modernity the salubrious effect such modernity might have on those who worked within the structures. The new factories were projects not just of aesthetic and economic import but also of social progressivism.

The Factory and the Worker

When the architect Warren Briggs wrote on "modern American factories" for the readership of *Architecture* in 1918, his approach was that of many building industry professionals of the time. After enumerating the improvements in lighting, ventilation, and sanitation seen in the "average" new American factory, Briggs claimed that to visit such a place was both a "pleasure" and an "inspiration," "for it will be found after a thorough inspection that the employee's life, instead of being, as in the past, a round of dirty drudgery, is really an ideal existence for those who have to work, as they spend their working hours in structures designed by skilled men and constructed in the most hygienic way known to modern science."[19] Well-designed surroundings were widely believed to have a positive influence on the "spirit and standards" of workers, offering employers returns in worker health and efficiency.[20]

This linking of an improved physical environment with workers' good spirits and health and the connection of these attainments with high productive standards and efficiency reflect the continuing embrace by factory builders of contemporary philosophies of management. The interest in "industrial relations" that had prompted building firms to organize the work of their own labor forces on the construction site now inspired them to study the situation of workers in the companies for which they erected factories. Here the builders turned not to ideas of systematic management but to a second trend in industrial administration: the institution of "worker welfare" policies that addressed the health, morale, and general well-being of industrial employees.

The earliest manifestations of an institutionalized concern with worker welfare in the United States had come in the housing, pension, and profit-sharing schemes of steel companies and railroads in the 1880s and 1890s. Firm owners intended these innovations to encourage company loyalty and reduce worker turnover while improving the financial circumstances of workers. As industries of all kinds consolidated towards the end of the century, new and larger populations came together to create workforces not all of which were easily managed with conventional, highly personalized shop-floor interventions. Worker-welfare programs offered an attractive management tool to a wide range of employers, allowing them to deter a growing movement towards labor militancy and to attract workers in competitive markets. University departments, government agencies, and trade organizations researched and disseminated information in this field of labor administration, while businesses established personnel departments that applied these findings to the recruitment and handling of labor. Among the most common provisions associated with worker welfare from 1900 onward were improvements to physical plants and programs for housing, health care, education, and recreation.[21]

Plans for improving industrial workers' comfort and safety inside the factory at the turn of the century brought improvements in factory lighting, heating, ventilation, and sanitation.[22] Prescriptive texts for factory owners frequently contained advice on topics ranging from the optimal size of windows to the most healthful number of toilets per building. Some innovations, such as air conditioning, were associated with contemporary ideas about physical health, and others, such as the choice of colors for factory walls, with ideas of workers' mental comfort.[23] Many large factory-building firms of the 1900s and 1910s that had integrated their services to include all phases of plant design and construction added the service of advising clients on the physical and moral well-being of factory employees.

The Aberthaw Construction Company, whose construction procedures are described in chapter 4, offers an example of a building company that took on the field of worker welfare. The firm's staff examined the physical nature of factory work. In a 1919 study conducted for the Seamless Rubber Company of Connecticut, Aberthaw's staff concluded that an outdated plant exacerbated the "unpleasant" nature of rubber making. They recommended a new facility designed to reduce the dust, damp, heat, and fumes created in rubber production (figs. 5.2, 5.3). Closely associating its buildings with a contented factory workforce, Aberthaw created pamphlets that advertised the company stores and doughnut bakeries included in the industrial plants it built. These accounts described the workers' bands and parades that greeted the opening of new facilities and even went so far as to cast the modernization of production processes as a secondary agenda

FIGURE 5.2. "Winter Calls No Halt, but It Compels a Blanket." Seamless Rubber Company plant, New Haven, Conn., under construction by the Aberthaw Construction Company, 1919–20. From *"Seamless: How Close Knit Cooperation Developed a Rubber Factory a Quarter Century in Advance of Its Time,"* brochure, Aberthaw Construction Company, ca. 1920, ACC Archives

among its clients. Of Seamless's new plant, completed in 1920, Aberthaw copywriters claimed, "There is much besides mechanism in the Seamless Rubber outfit. It is humanized first, and mechanized only in so far as is necessary to the directing of mutual good will toward unified accomplishment."[24] Cheerful photographs of smiling workers appeared beside images of the new factory building under construction by Aberthaw's forces.

Aberthaw's emphasis on the "human" element of modern factory production softened some of the sense of stark efficiency conveyed in the utilitarian factories' outward appearance and in the company's rhetoric about efficient factory operation. But Aberthaw's interest in workers' physical and emotional well-being fit well within its role as a source of modern industrial expertise. "Humanization" did not imply a lessening of organizational rigor for the factory owners who would buy Aberthaw's buildings; rather, it simply extended Aberthaw's expertise about factory design and operation to include workers' experiences of the modern plant. Aberthaw envisioned a paternalistic relationship between industrial employer and employee, and it offered advice on the subject to its clients. Once the building firm had included workers' experiences as a subject for its expert attention, it had little reason to stop at the factory door. Aberthaw, like other

FIGURE 5.3. "The Total Is One of Strength and Dignity." View of completed Seamless Rubber Company plant. *From "Seamless: How Close Knit Cooperation Developed a Rubber Factory"*

large factory construction firms of the day, also advised its clients on housing for factory employees.

In some instances Aberthaw's staff simply studied existing housing to assist in plant location; in other cases the firm erected entire "industrial villages." Among Aberthaw's largest such projects was a group of one hundred concrete homes it built in Donora, Pennsylvania, in 1912 for the American Steel and Iron Company.[25] Worker housing had by this time become a common subject for study by experts in industrial relations. Factory owners had supplied worker housing on a significant scale since New England mills began operation in the early nineteenth century, but by the early 1900s the subject had become one of almost scientific scrutiny for American industrialists. By 1910, most authorities in plant siting agreed that in order for a factory to have a dependable source of labor, it must be placed within walking distance or easy transport of adequate housing supplies. More specific investigations soon followed. In 1916 the Bureau of Labor Statistics worked with Harvard University's Department of Social Ethics to survey and interpret trends in American industrial housing, responding to industrial employers' feelings that "there are certain specific results, particularly in relation to the character and loyalty of the labor force, to be obtained by supplying an improved type of housing."[26] The study counted numbers and types of housing units supplied by more than two hundred companies in all regions of the country, and assessed such factors as how crowded various worker houses were and how "artistic" company towns were in appearance.[27]

In researching relocation and construction questions for their clients, Aberthaw's experts joined this quantifying trend. They ascertained the numbers and kinds of housing units near a proposed factory site and counted the numbers and "kinds" of workers—male or female, married or single—available. Then they compared these findings with the labor needs of the client company.[28] Aberthaw's staff saw important ramifications for this research. In 1920 the firm published a brochure written by Morton Tuttle, Aberthaw's general manager, titled "The Housing Problem in Its Relation to the Contentment of Labor."[29] In this tract Tuttle described the discontent that takes over the worker's household if his family is not comfortably and affordably housed. Tuttle emphasized that an industry could suffer greatly if it did not have adequate housing stocks available for its workers. The brochure attracted customers for Aberthaw's services and also provided arguments that industrial clients might take to government. Tuttle stated that a city or state should take responsibility for such housing if it wished to secure a strong industrial base.

Like gain sharing or improved job safety, reasonably priced, good-quality company-supplied housing in some ways constituted a benefit for factory employees, protecting them from crowded and often exploitative urban housing markets. But Tuttle's pamphlet does not frame the question of housing in the language of philanthropic impulse or social reform. Instead, Tuttle casts the average American worker as both ignorant and dangerous, to himself and to society in general: "Industrially, family discomfort is certain to work out in discontent with its direct consequences of low production and a state of mind which readily accepts the preaching of the red radical."[30] Putting aside the question of how extensive the presence of socialism or communism on American work sites may have been at this time, we can gather from Tuttle's formulations that he did not credit workers with any kind of political sophistication. Nor did he grant them the capacity to manage their own economic and family affairs in an acceptable way. Tuttle mentions that no worker could be expected to save money towards the purchase of a house because "that requires imagination." In short, Aberthaw's paternalistic stance towards factory workers could cut both ways, generating either progressively minded improvements to the workplace or condescending judgments that might restrict workers' control over their own lives.

Other services offered by Aberthaw also point to the equivocal nature of some modern management philosophies. For example, in helping Seamless Rubber design its new Connecticut plant, Aberthaw staff members recommended physical improvements to attract a different "class of workers" from the female and "unnaturalized Italian, Italian American, Russian, Pole, Jew and Syrian" male workers drawn to Seamless's existing plant.[31] We might also ask: If doughnut shops and cheerful lunchrooms were considered adequate to boost workers' morale, were larger improvements in wages or job mobility dismissed as unnecessary?[32] Some authorities in factory design seemed to take such a limited view of worker entitlement. Consultant Willard Case wrote in 1919,

> With the growth of our industries there fortunately followed apace an increasing perception of the value of properly housing our manufacturing plants. Necessity first furnished the inspiration of better factory buildings in answer to purely economic demands. The movement gained considerable impetus because of the business sagacity of those who experienced its benefits; it has been fostered in no mean way by the growth of the spirit of the "square deal" towards the rights of our workers.[33]

Case goes on to identify the contemporary factory building as no less than an expression of the "present day atmosphere of industrial freedom, joy in work, equality of all labor—brain and manual." He imputes to workers a pleasure in laboring for others' profits. This is an important facet of corporate paternalism that reads a pathology or anti-Americanism into workers' discontent.[34]

Aberthaw's attitudes towards workers seem to have fallen between the extremes of benevolence and constraint that the firm's advertising rhetoric and Case's claim, respectively, imply. The firm's factories did have attractive features and were carefully designed to ease working conditions. But Aberthaw's philosophies on worker hiring and housing, while not unusual for the day, suggest something of the tenuous authority factory workers had over their own work and personal circumstances. The modern factory building embodied a complex and shifting set of relations between those who built and managed the workplace and those who labored there.

The Industrial Response to the Functionalist Aesthetic

The multifaceted services that Aberthaw offered its clients encompassed a range of benefits: the practical advantages of the economically and scientifically constructed concrete factory building; aesthetic innovation; and means of social organization for industrial production. Other factory-building firms offered the same mix of services, but the popularity of the reinforced-concrete factory building cannot be fully explained by study of its suppliers. How did buyers of the new factories see these offerings? Which of these features was most important in drawing manufacturers and other industrialists to the new building type? How best can we understand the widespread demand for reinforced-concrete factory buildings after 1900?

A full response to these questions would require an investigation that lies outside the scope of this study. The complexities of industrial expansion and labor relations in the early twentieth century have long been recognized as deserving the attentions of entire historical disciplines, and patterns of factory buying can best be studied within these frameworks. Even to outline the issues involved in industry's changing architectural taste is no easy task. Industrialists rarely discussed the appearance of their physical plants in terms more specific than *modern, up-to-date, attractive, economical,* and *efficient.* More detailed descriptions appear in contracts between companies commissioning factory buildings and building or engineering firms, but these documents are generally highly technical and do not address the question of why buildings looked as they did. Manufacturers and mill owners did not explain why their new plants did not retain the rich brick detailing, terra-cotta mosaics, and clock towers popular in earlier decades, but we can begin to sketch a path for this inquiry by looking at the evidence to hand: the buildings themselves. Especially informative are cases in which we can compare multiple buildings owned by a single company, erected over time or for different purposes. One particularly telling example is the group of buildings erected by the Delaware, Lackawanna, and Western Railroad (DL&W) in 1909. All were built at the same time, on the same site, and with the same construction forces, yet each has a different visual character, showing a greater or lesser embrace of the functionalist aesthetic.

The DL&W was a wealthy corporation with rail lines, shipping facilities, and related coal and steel businesses throughout the northeastern United

States. In the 1880s, Scranton, Pennsylvania, served as a division point on the railroad's main line between Hoboken and Buffalo. By the turn of the century Scranton had become an important site of locomotive maintenance. As part of a major modernization program, the railroad began in 1907 to construct a 23-acre complex of reinforced-concrete "erecting," or repair, shops; foundries; blacksmith shops; gas production and oil storage buildings; and office and warehouse facilities. The new Scranton shops were designed by the DL&W's chief architect, Frank Nies, who had also designed the railroad's large passenger station and many signal towers in Scranton. Comparison of the appearances of these structures gives a sense of how the railroad perceived the architectural meanings of different materials and stylistic features.[35]

The building program at Scranton, largely completed by 1909, cost the DL&W more than $2 million. Efficient building function and visual effect, but not necessarily maximum economy, were paramount goals. While almost all of the 1909 structures were of similar construction, using reinforced-concrete-skeleton or steel-frame methods employed in repeating bays, the buildings bear differing amounts of ornamentation. They range from the utilitarian to the nostalgically ornate, depending on their function.[36]

The DL&W's office/warehouse building falls squarely in the middle of this spectrum. It is a three-story rectangular concrete-frame building 260 feet long and 60 feet wide. The building is brick-clad, has some brick detailing on its surface, and features cast-concrete details such as lintels, sills, and caps marking setbacks of columns. The office/warehouse building is of a simpler design than Nies's elaborate neoclassical passenger station, which stands two blocks away, and is more austere than the tile-roofed concrete signal towers that dot the tracks running into and out of Scranton. The building is more "elegant" than the locomotive-erecting shop that faces it across a local street.

The erecting shop, in which huge locomotives were disassembled and repaired, was built with steel framing to support a 55-foot-high open interior, through which heavy-duty cranes ran along the shop's 582-foot length. Although the interior of the building is largely uninterrupted, the exterior emphasized the regular spacing of steel columns in a repeating-bay scheme that looks much like that of the office/warehouse building. The erecting shop, however, presents a surface with very little detailing. It is brick-clad but lacks even the simple ornamentation and variation of its neighbor. It is more utilitarian in mien than the office/warehouse. The foundry building—400 feet long, 120 feet wide, and steel-framed with brick curtain walls—stood next to the erecting shops and had a similarly stark appearance.

A fourth structure, standing beside the office/warehouse building and connected to the erecting shops by tunnels and piping, held gas production equipment. This reinforced-concrete building generated gasses used for the operation of furnaces, foundries, and various machining operations on the premises. This two-story gas house, much smaller than the buildings around it, was designed to contain only three gas furnaces and a storage tank. It not only bears no cladding or brickwork of any kind but also displays on all surfaces, inside and out, the rough texture of the wooden boards from which its concrete forms had been built. It can be contrasted with Nies's signal towers; these also had exposed concrete surfaces, but the concrete was carefully brushed and hammered to produce a range of decorative effects. The gas house, in its simplicity and frank expression of materials, stands out as the most explicitly utilitarian of the 1909 buildings.

In the range of decoration and functionalism that Nies brought to the Scranton yards there is evidence of a stylistic order, a hierarchy of symbolic architectural forms. In the buildings through which passengers passed (passenger station) or that stood beside passenger-bearing tracks (signal towers), Nies turned to traditional historicizing motifs. In the office/warehouse building, where management did its work of administering the site and controlling inventories of supplies—both activities could be seen as types of mental labor—Nies provided a structure with a moderate level of ornament. Perhaps there was enough detail to convey the intellectual nature of the work being done inside without suggesting the lyricism of his granite-columned passenger station.

The erecting shop and foundry were also meant to be pleasing to the eye, with their regular shapes and warm-red brick cladding concealing the actual arrangement of inside space. However, they lacked the visual interest of the office/warehouse building, which would here have been regarded as excess. The productive work inside the erecting shop and foundry involved skill and seriousness of purpose, but unlike the mental labor conducted in the office building, it was largely divorced from other, older cultural enterprises, references to which comprise conventional architectural detailing. Nies felt no commitment to conventional decorative agendas. The gas house, a building that housed machines and only a handful of workers, referred in its design to little more than its function as a container.

The self-consciousness with which Nies and his supervisors approached building design can be confirmed by one crucial detail that appears in all the buildings of 1909. Each bears, on all sides facing thoroughfares or tracks that bore passenger trains, a large cast-concrete panel showing its name, and thereby its function. The deeply incised block letters are set inside a very simple incised border and announce that the passerby is seeing the "Erecting Shops," the "Foundry," the "Gas House," and so on. The signs no doubt provided guidance for suppliers or other visitors to the site, but their proud placement on the tops of the buildings' facades, rather than just at eye level, suggests that the function of each building formed part of its public identity. Each structure had a role to play in the conduct of the DL&W's business, and an aesthetic character to match that role.

To extrapolate from the DL&W's architectural example, we might say that for early-twentieth-century industries, an emphasis on architectural utility and standardization did not necessarily indicate a desire for corporate anonymity. The buildings on the Scranton site publicized the modern character of the railroad's work, and the uniform, utilitarian appearance of other reinforced-concrete industrial buildings publicized their owners' pride in their commitment to rationalized production. This brief discussion begins to sketch the aesthetic interest that functionalist building design held for American industrialists. It is intended to convey the important point that the meaning of this architecture to business people extended beyond a purely economic one into a sphere of public identity and aesthetic influence.

Conclusion

We have located several sources and implications for early-twentieth-century factory builders' appreciation of the functionalist architectural aesthetic. The builders sought to achieve a level of conventional artistic accomplishment with the simplified forms of reinforced concrete. To this end, they associated the repetitive design of the quickly erected reinforced-concrete factory building with beauty, harmony, and other traditional measures of "high" architecture. Their program found confirmation among forward-looking architects and critics, many of whom also praised the factories for their frank expression of commercial function. Builders and architects alike considered aesthetic value and advertising value to be fundamentally compatible.

Finding this agreement between the rhetoric of the factory builders and the modernist architects and critics, it is difficult to judge if the builders were seeking the approval of architects and critics, or if the architects and critics wished to lay intellectual claim to what was quickly becoming a popular new building style. Both probably took place to some extent, but of more importance may be the full nature of the consensus displayed by the two groups. Builders and observers of architecture were formulating new ideas of what was of value to American culture in the new century. The "realism" of economical construction methods and the creation of a positive commercial reputation were declared to be as important as any conventional artistic undertaking. Thus, the utilitarian architectural form was claimed to be as valid a design scheme as any historical building motif. This validation may have laid the groundwork for the functionalism that spread well beyond the industrial sector in following decades, persisting as a favorite American building style for high architecture and vernacular projects through most of the century.

The modernizing vision of factory builders also encompassed changes to the daily experiences of workers, some that would enhance the safety and comfort of factory employees, others that would limit their options on the job and in their personal lives. The work and rhetoric of the Aberthaw Construction Company in the area of worker welfare suggest that altruistic and self-serving intentions could coexist in such reform schemes for factory redesign. Modernity and true social progressivism were not coextensive for all early-twentieth-century factory builders.

The mixed social objectives displayed by factory builders imply a complexity in the attitudes of the industrialists to whom they sold their services. That the owners of reinforced-concrete factory buildings had concerns beyond simple economizing or technological expediency in choosing these structures is demonstrated in the variable styles chosen by one industrial company for a complex of utilitarian structures. Such variation could have had neither an economic nor a technological basis alone but rather shows a set of symbolic intentions for the modern industrial site as well. The functionalist public faces of industrial buildings arose from an architectural language different from that of older styles but no less self-conscious.

Conclusion

> For years we knew not how to build, and spent
> Our efforts all in vain, with good intent
> Rebuilding often. Now we have cement.
> —*Charles P. Stivers, 1914*

For those who designed, built, owned, and worked within concrete factory buildings in 1900 or 1930, the modernity of such structures was palpable. As architectural forms, technological artifacts, and products of emerging occupational jurisdictions and market forces, such buildings expressed departures from familiar practices. On all of these levels the buildings reflected a commitment to the large-scale, rationalized operations of industrial enterprise. But in each regard concrete factory buildings also reflected some traditional values, selected to achieve for engineers, builders, and industrialists a certain social dominance in the emerging industrial culture. Of primary importance in this telling is the blended nature of technical expertise about concrete—a combination of new knowledge and practices and the comprehensive, subjective approaches of "prescientific" construction methods. If we are to understand science and engineering as categories of thought distinct from older approaches to technical problem solving, it will have to be on the basis of their social organization, not through the identification of any clear epistemological trait such as "systemization" or "objectivity." Reinforced-concrete factory buildings of the early twentieth century reveal both the hybrid character of modern technical expertise and the profound implications of its social features well beyond the borders of scientific and engineering disciplines.

Technical experts in the new century shaped new bodies of knowledge and new techniques for industrial production with an inherently social understanding of how they would be used and by whom. When university instructors created testing techniques for concrete and taught those techniques to their students, they did so with an acute sensitivity to the conditions of commercial employment. Their awareness extended from the technical exigencies of field-based testing to the hiring and supervisory structures of the construction industry. The design of tests for concrete, testing instruments, and instructions for the use of those devices all reflect a hierarchical conception of expertise enacted on the building site. The inclusion of subjective judgments in much of the outwardly rationalized work of quality control for concrete gave tremendous power to materials instructors seeking to assure their own, and their students', high status in the commercial sphere. Codifications of testing and construction procedures, embedded in model and project specifications, solidified these occupational advantages in the geographically dispersed world of concrete construction.

A further category of social interaction, trust, figured largely in the formation of testing techniques for concrete. In ascribing to trained engineers good character and high moral standing, educators and those who hired the young engineers attached personal traits to the efficacious use of science. A test procedure meant nothing in the hands of an uncertified technician. In exchanges between building concerns—engineering firms, contractors, suppliers of cement or other materials—trust continued to be an issue. A scientific procedure or scientifically assessed product could not alone be said to bring about quality; only if the practitioner was trustworthy was a practice to be relied upon. For both academics and businessmen who worked with industrial materials, a danger was clear: when scientific know-how became something to sell on the open market, the mere invocation of science-based quality-control techniques could enlist customers. If neither university professors nor businessmen had felt that unreliable practitioners were "fooling" some people, they would not have worked so hard to delineate the difference between true and false authority. Their judgments about others in this regard may have been colored by their own economic and occupational interests, but even if that was the case—especially if that was the case—their efforts reveal the power that technical knowledge holds in our culture apart from its actual capacity to solve technical problems.

The consequences of this power after 1900 were many. An aura of technical authority could be misattributed. Sometimes quality-control techniques were inexpertly used, and in construction this misuse may well have led to lapses in safety and economy for those building, buying, or occupying concrete structures. But of greater consequence are the exclusionary features of this phenomenon. The mantle of scientific authority was not something that arose directly from the way a person handled materials and instruments. Rather, it came with admission to university engineering programs or through the approval of professional and trade organizations. This in itself was a type of quality control, no doubt assuring that few unqualified practitioners entered the ranks of industrial employment. But in indicating that the tester, not the test, was what assured quality, established professionals created a gate through which many people—women, people of color, self-taught engineers—could not pass.

In both academic and commercial contexts, we have seen that notions of what constituted modern technical abilities in no way displaced all conventional valuations of technical acumen. On one level, the standardization of work processes after 1900 comes to seem a remarkably complex phenomenon, steeped in such "irrational" matters as subjectivity and character. However, by instituting highly variable technical practices (those operations based on judgment or character), technical standardization, as seen in the case of concrete construction, brought about a uniformity of social experience (almost assuring the occupational privilege of college-trained testers and the low status of manual laborers). This was a well-planned application of ambiguity.

On another level, for engineering faculty and industry leaders involved in the creation of standardized products and protocols, an "old-fashioned" style of technical problem solving, based on integrated and cumulative bodies of information, was both necessary and esteemed—as long as it was applied towards modern industrial purposes such as standardization. The integrated knowledge held by most building artisans trained in traditional apprenticeship systems, such as masonry or carpentry, was not of value to modern employers. Obviously, managerial strategies of divided labor and low wages could accommodate only so many highly skilled individuals. But we must recognize that modern managers in the concrete construction business were denying the autonomy and possibilities of upward mobility embedded in guild-style occupations to the majority of workers. It was a choice consciously made. The social alterations that came with economizing methods of concrete construction presented no moral dilemma for the technical experts and building firm owners.

Using that denial as an analytical starting point, we can detect a commonality of interest between those who developed scientific features of concrete use and those who built concrete buildings. Both groups prejudged members of the workforce on such "inherited" traits as gender and ethnicity. This is certainly not an unexpected pattern for the early twentieth century, but it can lead us to an important conclusion about industrial science. The use of technical expertise by industry after 1900 was not a matter of "business calling" and "science answering," the industrialists creating a demand that the academics and their progeny could then fill. Scientific disciplines promoted many of the technical advancements used by industry, but more important, those who made their living in technical fields shared a social and cultural milieu with business owners and operators. The techniques developed for carrying science to the industrial context embodied and encouraged the social divisions that both groups believed in. Yes, engineers solved problems for industry. They permitted the erection of stronger, cheaper buildings. But that was only part of the "problem" that industry wanted solved. The other was the maintenance of structures of occupational and social mobility drawn along lines of gender and class and race—also seen as a problem by technical experts in this era.

Because this book posits an adversarial relationship between those who possessed expert knowledge early in the twentieth century and those who did not, a final point needs to be made about how knowledge becomes power. The social meanings granted to information are often shared across groups of differing social standing—not visited upon an unwilling audience by inexplicably endowed elites. Pleas for modernity, trust in instruments, and the idea that technical knowledge is cumulative configure the writing of carpentry and masonry trades in this period. An editor of one carpenters' trade journal recommended mechanized measurement in all work: "I defy any mechanic, no matter how clever or experienced, to say or prove that his placement of timber by eyesight . . . is exact, whereas the level and plumb are correct in their application, for true instruments cannot, if properly applied, ever err or prove incorrect."[1] Advertisements for correspondence courses and advice manuals aimed at construction laborers reiterated that knowledge brings social mobility. The International Correspondence School of Scranton, Pennsylvania, a famously successful source of technical training for laborers, told readers in one 1914 advertisement that "today it's a battle of wits—and brains win! Muscle and brawn don't count as much as they used to!"[2] All manner of encyclopedia and consulting services promised to augment the carpenter's or bricklayer's "practical experience" with the more lucrative techniques of reading blueprints or checking specifications that "any big contractor knows."[3]

Elite occupations and workers in the building industry shared a second important notion: that different people simply possess different levels of innate technical ability; a stratified workplace and society made sense to both groups of practitioners. The general secretary of the Carpenters' Union wrote in 1916 that "the qualifications and capabilities of the average human being are so diversified" that choosing one's profession is a major challenge in the life of each American youth. Judgments about the abilities of fellow workers were sometimes associated with race or ethnicity. In 1916, carpenters urged passage of stricter immigration laws to keep "destitute laborers" from entering the country; bricklayers in Philadelphia excluded black workers from their union's local at the end of the nineteenth century.[4] Certainly the complete absence of women from skilled building trades in America was no more accidental than the gender biases exhibited by engineers on this subject. We must note that many of the major trade unions worked hard to address social inequities of race even before 1900 and to accommodate immigrant workers with training programs and other types of assistance; workers may in many instances have held more inclusive ideas about American culture than did elite or wealthy citizens. But we must be cautious about ascribing completely separate cultures to workers and to the managers and experts who made up the growing white-collar sector of the construction industry. Clearly, the immense question of whether and how capitalism perpetuates social inequities cannot be answered if we imagine some ideological firewall between social groups. The limited radicalism of American trade organizations has long been understood to be a complicated historical question that implies such unifying ideas about democracy across classes in this country.[5]

    The history of concrete use in the United States allows us to follow the role of technical acumen in enacting one set of capitalist work relations. Since 1930 the concrete industry has continued to develop and implement means of eliminating discretionary elements of concrete use in the field. First, dry ingredients were premixed and packaged at cement plants (the "ready-mix" concretes), removing the tasks of proportioning and mixing from the work site. Then, in the late 1930s, the practical difficulties of controlling water/cement ratios were overcome to a large degree by the introduction of "transit-mix" concrete, in which all ingredients were premixed and trucked to a construction site in revolving-drum trucks. Later decades saw the introduction of extensive systems of precast concrete elements that further removed responsibility for concrete fabrication from the work site. However much they may have speeded up concrete construction or lowered the its cost, these changes cannot be regarded simply as technological advances but must be recognized as markers of the still modernizing world of productive labor.

As symbolic forms, and as direct evidence of their creators' activities, reinforced-concrete factory buildings stood for much of the twentieth century as heralds of this particular type of modernity. Many concrete factory buildings now decline into ruin as a service economy replaces the manufacturing base in America, but where demand exists, some structures have found new uses as sites of light manufacturing or retailing. Perhaps most interesting is the late-twentieth-century wave of fashion that has brought the utilitarian factory buildings back into an architectural classification of "good taste." In larger cities, highly functionalist exposed-frame concrete factories are being renovated for conversion into expensive residential units.

The rehabilitation of factory buildings as urban status symbols dovetails neatly with the continuing allure that high levels of productivity hold for many Americans, even as that productivity shifts from its industrial manifestations to seemingly dematerialized information-age expressions. To root that taste for large-scale production, and its partner, large-scale consumption, in the social relations that accompany corporate growth is to alert oneself to the mixed consequences that industrial expansion has had for different groups of workers in the twentieth century—and, we might add, to see the material consequences of information that becomes commodified. As the engineering curricula, instruments, and materials standards and specifications of the early twentieth century translated knowledge into occupational and commercial advantage, so contemporary representations of knowledge in electronic form may carry particular social effects—disseminating "best practice" across time and space with remarkable velocity. In the form of texts or databases, the products of science and engineering may seem insubstantial; but their consequences are concrete, in every sense of the word.

Notes

Abbreviations

ACC Archives	Files of the Aberthaw Construction Company, Billerica, Massachusetts
ACI *Proc.*	*Proceedings of the American Concrete Institute*
ASTM *Proc.*	*Proceedings of the American Society for Testing Materials*
ASTM *Yearbook*	*Yearbook of the American Society for Testing Materials*
ISU-MP	Iowa State University Archives, Marston Papers
ISU-GP	Iowa State University Archives, Gilkey Papers
JACI	*Journal of the American Concrete Institute*
NBS *AR*	*Annual Report of the National Bureau of Standards*
SPEE *Proc.*	*Proceedings of the Society for the Promotion of Engineering Education*
UIA	University of Illinois Archives
UPA	University of Pennsylvania Archives

A note of explanation is required regarding two of the journals used in this study. *Cement Age,* through the first two decades of the twentieth century, repeatedly altered its volume numbering as it absorbed other periodicals; volumes are not numbered consecutively for this reason. Similarly, the publications of the Society for the Promotion of Engineering Education in this period follow a rather erratic system of titling (the society's *Bulletin* and *Proceedings* were at times separate publications, at times combined, and were finally subsumed in 1916 by *Engineering Education*) and volume numbering. The reader may find helpful Matthew Elias Zaret, "An Historical Study of the Development of the American Society for Engineering Education" (Ph.D. diss., New York University, 1967), 81–83.

Preface

1. On the notion that historians have not explicated industrial growth as a cultural choice in America, see David Montgomery, *Workers' Control in America,* 2d ed. (Cambridge: Cambridge University Press, 1996 [1979]), 1–4.

2. This was the case in the Ames (Iowa), Champaign-Urbana (Ill.), and Philadelphia areas—all locations of large university engineering schools. See, for example, Correspondence Files, Towne Scientific School, 1900–1920, UPA, Record Group UPD 2.2; and "Engineering Alumni Reports" 1911–14, and Alumni Files, UPA.

INTRODUCTION. Science and Commerce

Epigraph: J. A. L. Waddell, "Address to the Society for the Promotion of Engineering Education," SPEE *Proc.* 23 (1915): 215.

1. Two studies have been most influential for this project. David Noble's unparalleled *America by Design: Science, Technology, and the Rise of Corporate Capitalism* (New York: Oxford University Press, 1977) establishes the inherently social and self-consciously hierarchical nature of technical practice in the industrial context. Geoffrey Bowker, in *Science on the Run: Information Management and Industrial Geophysics at Schlumberger, 1920–1940* (Cambridge, Mass.: MIT Press, 1994), embeds the creation of scientific information and authority in the competitive conditions of industrial employment (which derive, he shows, from political, racial, and cultural ideologies). For extensive discussion of relevant literature in the history of industry, technology, and building, see "Bibliographic Essay."

2. Leonard S. Reich, *The Making of American Industrial Research: Science and Business at GE and Bell, 1876–1926* (Cambridge: Cambridge University Press, 1985); George Wise, *Willis R. Whitney: General Electric and the Origins of U.S. Industrial Research* (New York: Columbia University Press, 1985); Stuart Leslie, *Boss Kettering: Wizard of General Motors* (New York: Columbia University Press, 1983); David A. Hounshell and John K. Smith, *Science and Corporate Strategy: DuPont R&D, 1902–1980* (Cambridge: Cambridge University Press, 1988); Paul Israel, *From Machine Shop to Industrial Laboratory: Telegraphy and the Changing Context of American Invention, 1830–1920* (Baltimore: Johns Hopkins University Press, 1992).

3. Despite its persistence as a managerial tool and topic of prescriptive literature, science-based quality control has received little attention from historians of business, labor, and technology. Important exceptions include Bruce Seely, *Building the American Highway System* (Philadelphia: Temple University Press, 1987); Noble, *America by Design*, 69–83; David C. Mowery and Nathan Rosenberg, *Technology and the Pursuit of Economic Growth* (Cambridge: Cambridge University Press, 1989), 35–58; Bruce Sinclair, *A Centennial History of the American Society of Mechanical Engineers, 1880–1980* (Toronto: University of Toronto Press, 1980), 55–57, 146–57; and Virginia P. Dawson, "Knowledge Is Power: E. G. Bailey and the Invention and Marketing of the Bailey Boiler Meter," *Technology and Culture* 37 (1996): 493–526. David Hounshell's *From the American System to Mass Production, 1800 to 1932* (Baltimore: Johns Hopkins University Press, 1984) and Ken Alder's

Engineering the Revolution: Arms and Enlightenment in France, 1763–1815 (Princeton: Princeton University Press, 1997) describe early approaches to industrial quality control. On the creation of scientific standards for commercial use, see Simon Schaffer, "Accurate Measurement Is an English Science" (135–72), and Graeme J. N. Gooday, "The Morals of Energy Metering: Constructing and Deconstructing the Precision of the Victorian Electrical Engineer's Ammeter and Voltmeter" (239–82), both in *The Values of Precision,* ed. M. Norton Wise (Princeton: Princeton University Press, 1995); and Theodore M. Porter, *Trust in Numbers* (Princeton: Princeton University Press, 1995).

4. Halbert P. Gillette and George S. Hill, *Concrete Construction Methods and Cost* (New York: Myron C. Clark, 1908), 4–5, 477. See also Sanford E. Thompson and William O. Lichtner, "Scientific Methods in Construction," *Proceedings of the Engineers Society of Western Pennsylvania,* 1916, 434–65.

5. On the division of labor in inspection, see Michael Burawoy, *Manufacturing Consent: Changes in the Labor Process under Monopoly Capitalism* (Chicago: University of Chicago Press, 1979), and Robert B. Gordon, "Who Turned the Mechanical Ideal into Reality?" *Technology and Culture* 29 (1988): 744–78.

6. Andrew Abbott, *The System of Professions: An Essay on the Division of Expert Labor* (Chicago: University of Chicago Press, 1988); Sinclair, *Centennial History,* especially chap. 5, "The Arts Are Full of Reckless Things"; Stuart Shapiro, "Degrees of Freedom: The Interaction of Standards of Practice and Engineering Judgment," *Science, Technology, and Human Values* 22 (1997): 291–93.

7. For parallel cases, see JoAnne Brown, *The Definition of a Profession: The Authority of Metaphor in the History of Intelligence Testing, 1890–1930* (Princeton: Princeton University Press, 1992), and Peter Novick, *That Noble Dream: The "Objectivity Question" and the American Historical Profession* (Cambridge: Cambridge University Press, 1988).

8. Thorstein Veblen, *The Higher Learning in America: A Memorandum on the Conduct of Universities by Business Men* (1918; reprinted New York: Hill and Wang, 1957); Thomas L. Haskell, "Professionalism *versus* Capitalism: R. H. Tawney, Emile Durkheim, and C. S. Peirce on the Disinterestedness of Professional Communities," in *The Authority of Experts,* ed. Thomas L. Haskell (Bloomington: Indiana University Press, 1984), 191–206.

9. Noble, *America by Design,* 257–77. See also Sinclair, *Centennial History.*

10. For a typical expression of this emphasis on morality in technical practice, see Charles B. Dudley, "The Testing Engineer," SPEE *Proc.* 13 (1906): 233–51. For Dudley the implicit antagonism between tester and producer, or between competing producers using the services of testers, virtually guaranteed more rigorous scientific practice than one would find in the isolated world of "pure" research on materials. See also Michael Aaron Dennis, "Accounting for Research: New Histories of Corporate Laboratories and the Social History of American Science," *Social*

Studies of Science 17 (1987): 471–518; Steven Shapin and Simon Schaffer, *Leviathan and the Air-Pump: Hobbes, Boyle, and the Experimental Life* (Princeton: Princeton University Press, 1985); and Lorraine Daston and Peter L. Galison, "The Image of Objectivity," *Representations* 40 (1992): 81–128. On the historical connection of moral virtue and the reliance on faculties of judgment in science, see Peter L. Galison, "Judgment against Objectivity," in *Picturing Science, Presenting Art*, ed. Peter L. Galison and Caroline A. Jones (London: Routledge, 1998), 327–59.

11. A. Marston, Iowa State College School of Engineering, "The Engineer's Responsibility to the State," ms., 24 June 1931, ISU-MP, Record Series 11/1/11, Box 5, n.p.

12. Robert Kohler articulates the importance of "moral economies" in scientific domains, building on E. P. Thompson's conception of moral economies "that regulate authority relations and access to the means of production and rewards for achievement." Robert Kohler, *Lords of the Fly: Drosophila Genetics and the Experimental Life* (Chicago: University of Chicago Press, 1994), 3, 6.

13. Samuel Krislov, *How Nations Choose Product Standards and Standards Change Nations* (Pittsburgh: University of Pittsburgh Press, 1997). In later years, the ASTM became known as the American Society for Testing and Materials.

14. Natural cements established a foothold among engineers in the United States after 1820 when Canvass White, an engineer of the Erie Canal, patented a hydraulic cement (one able to harden under water) made from limestones found in New York State. Competing manufacturers in New York, Pennsylvania, and elsewhere on the eastern seaboard exploited other deposits, bringing cement into increasing popularity for canals and other masonry projects. Portland cements, manufactured in large quantities in England after 1850 and the United States after about 1880, were based on carefully proportioned combinations of clay and limestone, burned at higher temperatures than natural cements. They provided a readily available alternative to natural cements in geographical areas lacking natural cement deposits. Because their chemical composition could be precisely controlled, Portland cements were also generally of more predictable behavior than natural cements. See Thomas F. Hahn and Emory L. Kemp, *Cement Mills along the Potomac River,* Institute for the History of Technology and Industrial Archaeology Monograph Series, vol. 2, no. 1 (1994): 7–21; Jasper O. Draffin, "A Brief History of Lime, Cement, Concrete, and Reinforced Concrete," *Journal of the Western Society of Engineers* 48 (1943): 8–12, reprinted in *A Selection of Historic American Papers on Concrete, 1876–1926* ed. Howard Newlon Jr. (Detroit: American Concrete Institute, 1976).

15. Draffin, "Brief History," 8–9; William B. Coney and Barbara Posadas, "Concrete in Illinois: Its History and Preservation," Illinois Preservation Series of the Illinois Historic Preservation Agency, no. 8 (n.d.), 5.

16. Cecil Elliott, *Technics and Architecture: The Development of Materials and Systems for Building* (Cambridge, Mass.: MIT Press, 1992), 149–87; Carl Condit, *Amer-*

ican Building Art (Chicago: University of Chicago Press, 1968), 168; James C. Mabry, "Regulation, Industry Structure, and Competitiveness in the U.S. Portland Cement Industry," *Business and Economic History* 27 (1998): 402–12.

17. In 1854 William Wilkinson, an English plasterer, patented a type of poured-concrete slab in which inlaid wire cables—bought used, from mines—provided reinforcement. A significant understanding of reinforcing techniques is seen in methods devised by Josef Monier and François Coignet in 1867 and exhibited at the Paris Exposition of that year. See Condit, *American Building Art,* 168–69. For additional information on early patents for reinforced concrete, see also Draffin, "Brief History," 29; Henry J. Cowan, *The Master Builders,* vol. 1, *Science and Building: Structural and Environmental Design in the Nineteenth and Twentieth Centuries* (New York: John Wiley and Sons, 1977), 31; and Elliott, *Technics and Architecture,* 167–69.

18. Ward demonstrated that his house was virtually fireproof—it contained almost no wood—and that its parlor floor could withstand 26 tons of weight. Condit, *American Building Art,* 169–70. See also Elliott, *Technics and Architecture,* 173–74.

19. Cowan, *Master Builders,* 36.

20. Gwenael Delhumeau, "Hennebique: Les architects et la concurrence," in *Culture constructive,* ed. Patrik Bardou et al. (Paris: Editions Parentheses, 1992), 15–25.

21. In 1877 Hyatt published his test results in "An Account of Some Experiments with Portland-Cement-Concrete Combined with Iron, as a Building Material with Reference to Economy of Metal in Construction, and for Security against Fire in the Making of Roofs, Floors, and Walking Surfaces," reprinted in *Selection of Historic American Papers.* Hyatt advocated using bent and deformed reinforcing bars to achieve more precisely controlled and more complete interactions between concrete and reinforcement than had previously been possible. He specified longitudinal rods and circumferential reinforcing hoops for columns as well, greatly increasing the ability of columns to resist crushing and buckling. See Draffin, "Brief History," 30–31; Cowan, *Master Builders,* 36; Condit, *American Building Art,* 171; and Elliott, *Technics and Architecture,* 122.

22. Draffin, "Brief History," 32.

23. Ernest Ransome was the son of Frederick Ransome, who had contributed to the development of the British concrete industry. The younger Ransome recognized concrete to be a less expensive alternative to iron as a structural element. His important early concrete buildings included the California Academy of Sciences and the Museum of Stanford University. The Stanford building survived the San Francisco earthquake of 1906. Ernest Ransome, "New Developments in Unit Work Using a Structural Concrete Frame and Poured Slab," *Cement Age* 12 (March 1911): 130. See also Cowan, *Master Builders,* 36; Draffin, "Brief History," 31; Condit, *American Building Art,* 172; and Ernest L. Ransome, "Reminiscence," in *Rein-*

forced Concrete Buildings (New York: McGraw-Hill, 1912), reprinted in *Selection of Historic American Papers,* 285–305.

24. Elliott, *Technics and Architecture,* 176; Walter Mueller, "Reinforced Concrete Construction," *Cement Age* 3 (October 1906): 321.

25. Ransome, "Reminiscence," 301.

26. Coney and Posadas, "Concrete in Illinois," 4.

27. Ross F. Tucker, "The Progress and Logical Design of Reinforced Concrete," *Cement Age* 3 (September 1906): 230–31.

28. William A. Radford, *Cement and How to Use It* (Chicago: Radford Architectural Co., 1919), 14–15; Portland Cement Association, *Portland Cement Association Reference Book for Editors 1925* (Chicago: Portland Cement Association, 1925), 17.

29. Coney and Posadas, "Concrete in Illinois," 4; *Engineering News,* 6 November 1902, 385; Condit, *American Building Art,* 158.

CHAPTER ONE. Concrete Testing

Epigraph: Charles B. Dudley, "The Testing Engineer," SPEE *Proc.* 13 (1906): 249.

1. William E. Wickenden, "A Comparative Study of Engineering Education in the U.S. and Europe," in *Report of the Investigation of Engineering Education* (Pittsburgh: SPEE, 1930), 72.

2. Edwin Layton, "Mirror-Image Twins: The Communities of Science and Technology in Nineteenth-Century America," *Technology and Culture* 12 (1971): 567–68. Layton's observations apply to a relatively narrow portion of the profession: the leading edge of research, rather than the broad range of activity in the commercial sector.

3. This impression is encouraged by the historical conception that a "clash of cultures" occurred between shop- and college-trained engineers around the turn of the century. See Monte Calvert, *The Mechanical Engineer in America, 1830–1910: Professional Cultures in Conflict* (Baltimore: Johns Hopkins Press, 1967), 3–40, and Layton, "Mirror-Image Twins."

4. Lorraine Daston and Peter Galison, "The Image of Objectivity," *Representations* 40 (1992): 118.

5. The early-twentieth-century university departments that specialized in testing for industry did not suffer from the circumstances described by David Noble, in which "schools lagged seriously behind the rapidly changing needs of the science-based industries." Nor were the academic materials scientists at a disadvantage Noble describes: "The efforts of the engineering educators to meet the demands of their profession, academic status, and employers of engineers were never wholly satisfactory." David Noble, *America by Design: Science, Technology, and the Rise of Corporate Capitalism* (New York: Oxford University Press, 1977), 28–29.

6. See Geoffrey Cantor, "The Rhetoric of Experiment," in *The Uses of Experiment*, ed. David Gooding, Trevor Pinch, and Simon Schaffer (Cambridge: Cambridge University Press, 1989), 159–80.

7. The consideration of "mental habits" is central to much of the advice given by Clement C. Williams, dean of the College of Engineering at the State University of Iowa, in *Building an Engineering Career* (New York: McGraw-Hill, 1934). Noble, in *America by Design,* 200, describes elite conceptions of administrative work expressed by members of SPEE.

8. John Rae, "The Application of Science to Industry," in *The Organization of Knowledge in Modern America, 1860–1920,* ed. Alexandra Oleson and John Voss (Baltimore: Johns Hopkins University Press, 1979), 249–52; Leonard S. Reich, *The Making of American Industrial Research: Science and Business at GE and Bell, 1876–1926* (Cambridge: Cambridge University Press, 1985), 16–18; Wickenden, "Comparative Study," 64; Terry Reynolds, "The Education of Engineers in America before the Morrill Act of 1862," *History of Education Quarterly* 32 (Winter 1992): 459–82 (I am indebted to Bruce Seely for bringing this article to my attention). See also Daniel Calhoun, *The American Civil Engineer: Origins and Conflict* (Cambridge, Mass.: Technology Press, MIT; distributed by Harvard University Press, 1960), 16–20. Useful surveys of nineteenth-century American engineering education appear in Noble, *America by Design,* and Calvert, *Mechanical Engineer in America.*

9. Wickenden, "Comparative Study," 68–70.

10. Bruce Seely, "Research, Engineering, and Science in American Engineering Colleges, 1900–1960," *Technology and Culture* 34 (1993): 345.

11. Anson Marston, "Engineering Research in Land Grant Colleges," ms. from Marston to Department of Commerce, 21 November 1928, ISU-MP, Record Series 11/1/01/3, Box 5, 12; Seely, "Research, Engineering, and Science," 350–51.

12. David C. Mowery and Nathan Rosenberg, *Technology and the Pursuit of Economic Growth* (Cambridge: Cambridge University Press, 1989), 27–34. Also Reich, *Making of American Industrial Research,* 28–29, and Rae, "Application of Science to Industry," 249.

13. Mowery and Rosenberg, *Technology and the Pursuit of Economic Growth,* 79.

14. Jasper O. Draffin, "A Century of American Textbooks in Mechanics and Resistance of Materials," *Journal of Engineering Education* 36 (1945): 332–33.

15. Layton, "Mirror-Image Twins," 570–75.

16. Noble, *America by Design,* 185–87.

17. Frank McKibben, "The Design, Equipment, and Operation of University Testing Laboratories," *Engineering Education* 20 (1912): 145; Clarence A. Martin, "Courses in Architecture as Offered in American Universities, December 27, 1913," table, UIA, Record Group 12/2/1, Fine and Applied Art: Architecture 1910–1959, Box 2, File: "Association of Collegiate Schools of Architecture"; John G. D. Mack, "Trades Training for Non-Technically Educated Men," SPEE *Proc.* 9 (1901): 310; Her-

bert J. Gilkey, "What Departments Teach Mechanics?" typescript of talk presented to SPEE, Cambridge, Mass., 28 June 1937, ISU-GP, Record Series 11/7/11, Box 2.

18. W. Ross Yates, *Lehigh University: A History of Education in Engineering, Business, and the Human Condition* (Bethlehem, Pa.: Lehigh University, 1992), 124.

19. McKibben, "Design, Equipment, and Operation," 140–48; Duff A. Abrams, "The Structural Materials Research Laboratory at Lewis Institute, Chicago," *Thirtieth Annual Report of the Illinois Society of Engineers and Surveyors*, 1911, 88; Yates, *Lehigh University*, 124.

20. Ira O. Baker and Everett E. King, *A History of the College of Engineering of the University of Illinois, 1868–1945, Part I* (Urbana: University of Illinois, 1946), 394–95.

21. "A Tribute to Arthur Newell Talbot," *University of Illinois Bulletin* 35 (1938): 15–17; Baker and King, *History of the College of Engineering*, 402–3.

22. Baker and King, *History of the College of Engineering*, 403–6; Charles E. Rosenberg, "Science, Technology, and Economic Growth: The Case of the Agricultural Experiment Station Scientist, 1875–1914," in *No Other Gods* (Baltimore: Johns Hopkins Press, 1961), 162–65.

23. James W. Phillips, ed., *Arthur Newell Talbot: Proceedings of a Conference to Honor TAM's First Department Head and His Family, April 1994* (Champaign: University of Illinois, 1994).

24. Herbert J. Gilkey, "General Description of the Work Offered and Methods Followed in the Department of Theoretical and Applied Mechanics," ms., ISU-GP, Record Series 11/7/1, Box 1, n.p.

25. Anson Marston to President E. W. Stanton, 5 May 1903, ISU-MP, Record Series 11/1/11, Box 1.

26. On Marston's participation in the "Good Roads" movement, see Bruce Seely, *Building the American Highway System* (Philadelphia: Temple University Press, 1987), 111–12. On the creation of ISC's engineering facilities, see *Bulletin of the Iowa State College Division of Engineering* 6 (1908): n.p.; C. S. Nichols, "Iowa State College of Agriculture and Mechanical Arts Directory of Graduates of the Division of Engineering," 1912, ISU-MP, Record Series 11/1/11, Box 7, 41–43.

27. Herbert Gilkey, "Correspondence Files," ISU-GP, Record Series 11/7/11, Box 3; Erik Lokensgard, "Formative Influences of Engineering Extension on Industrial Education at Iowa State College," Ph.D. dissertation, Iowa State University, 1986.

28. "University of Pennsylvania: Dedication of the New Engineering Building, October 19, 1906," brochure, UPA, Record Group UPG 35.631, 1906, n.p.

29. Lesley never completed his studies but received an honorary bachelor of science degree from the University of Pennsylvania in 1871.

30. "The New Engineering Building and Equipment of the University of Pennsylvania," *Proceedings of the Engineers Club of Philadelphia* 24 (1907): 16–19.

31. The activities of Marburg, Berry, and other members of Penn's engineer-

ing departments between 1900 and 1920 are detailed in correspondence files held in the university's archives. See UPA, Record Group UPD 2.1, Boxes 1 and 2, General Correspondence Files. Berry became known for his extensometer, a portable gauge used to measure electronically the deformation of concrete under strain. The Berry gauge was still widely used in the 1940s. Irving Cowdrey and Ralph Adams, *Materials Testing Theory and Practice* (New York: John Wiley and Sons, 1925), 23; Herbert Gilkey, Glenn Murphy, and Elmer O. Bergman, *Materials Testing* (New York: McGraw-Hill, 1941), 17.

32. "Engineering Alumni Reports," 1911–14, Alumni Files, UPA; alumni records of the University of Pennsylvania School of Engineering, 1900–1918, UPA; C. S. Nichols, "Some Statistics of the Iowa State College Engineering Graduates," *Iowa Engineer* 11 (1911): 333–45; Anson Marston, "What Does the Engineering College Expect of the Construction Industry?" talk presented to the Construction Division of the ASCE, 20 January 1927, n.p.; Ernest McCollough, *Engineering as a Vocation* (New York: David Williams Co., 1912), 112–13; Leonard C. Wason, "The Problems of the Contractor," ACI *Proc.* 11 (1914): 372; George Wise, *Willis R. Whitney: General Electric and the Origins of U.S. Industrial Research* (New York: Columbia University Press, 1985), 172–77.

33. On the work of German universities for industry, see Stephen P. Timoshenko, *History of Strength of Materials: With a Brief Account of the History of Theory of Elasticity and Theory of Structures* (New York: Dover Publications, 1953), 130–31, and J. A. L. Waddell "Address to the Society for the Promotion of Engineering Education," SPEE *Proc.* 23 (1915): 207–19. The number of American students studying in Germany peaked in the mid-1890s. Noble, *America by Design*, 131. See also Wickenden, "Comparative Study," 68, and Lawrence R. Veysey, *The Emergence of the American University* (Chicago: University of Chicago Press, 1965), 121–50.

34. Wickenden, "Comparative Study," 70–71; Reich, *Making of American Industrial Research*, 23–24; Rae, "Application of Science to Industry," 255.

35. McKibben, "Design, Equipment, and Operation," 147–49; Virgil R. Fleming, "Some Notes on Conducting Tests in Materials-Testing at the University of Illinois," SPEE *Proc.* 20 (1912): 252. On the extent and specificity of materials exercises at Iowa State College, see files of ISU-MP and ISU-GP, in which many course syllabi are preserved.

36. See, for example, laboratory instruction manual for a 1904 course in testing materials at the University of Illinois, Department of Theoretical and Applied Mechanics, UIA, Record Group 11/11/22, and G. Saxton Thompson, "Testing of Engineering Materials," SPEE *Proc.* 20 (1912): 196–97. Also W. K. Hatt, "A Laboratory Course in Testing Materials of Construction," SPEE *Proc.* 13 (1906): 254.

37. Thompson, "Testing of Engineering Materials," 202; Anson Marston, "Abstract of Paper: Field Work in Civil Engineering at Iowa State College," undated, ISU-MP, Record Series 11/1/01/3, Box 4.

38. "Materials Laboratory," syllabus, ISU-MP, Record Series 11/7/11, Box 1, n.p. See also Gilkey et al., *Materials Testing*, 108.

39. Cowdrey and Adams, *Materials Testing Theory and Practice*, 16.

40. McKibben, "Design, Equipment, and Operation," 143; Edwin Marburg to Riehle Testing Machine Company, 24 May 1905, UPA, Record Group UPD 2.2, Box 1, File: "Marburg Correspondence." See also Hatt, "Laboratory Course," 256–57.

41. Cowdrey and Adams, *Materials Testing Theory and Practice*, 92–93.

42. "Equipment Cement Laboratory Preliminary Estimate," University of Pennsylvania Civil Engineering School, UPA, Record Group UPD 2.2, Box 1, File: "Engineering Equipment January 1904–1906." By the 1930s the flow table was more popular for classroom use than in the construction field itself. See also Gilkey et al., *Materials Testing*, 86, and laboratory descriptions throughout SPEE *Proc.* 20 (1912).

43. Untitled memos, UPA, Record Group UPD 2.2, Box 1, File: "General Correspondence January 1909–January 1912"; "O.U. Miracle Award," press release, June 1908, ISU-MP, Record Series 11/1/11, Box 7. See also Charles Riborg Mann, "A Study of Engineering Education," Joint Committee on Engineering Education of the National Engineering Societies (Carnegie Foundation) Bulletin no. 11 (1918): 23. Such practicality was also part of a trend towards specialization in American technical education at the turn of the century. Science faculty at this time were urging students to choose narrower research topics for their theses, to assure exhaustive treatment. W. G. Hale, "The Doctor's Dissertation," *American Association of Universities Journal*, 1902, 18, cited in Veysey, *Emergence of the American University*, 143.

44. Hatt, "Laboratory Course," 255.

45. J. F. Rhodes, Lake Shore and Michigan Southern Railway, to H. C. Berry, 19 December 1910, and H. C. Berry to J. F. Rhodes, 31 January 1911, UPA, Record Group UPD 2.2, Box 1, File: "Engineering Correspondence June 1909–January 1912."

46. "ISC Division of Engineering 1904," brochure, Dean's File "Brochures and Periodicals," ISU-MP, Record Series 11/1/0/3.

47. Daniel W. Mead, *Contracts, Specifications, and Engineering Relations* (1916; reprinted New York: McGraw-Hill, 1933), 6.

48. F. P. Spalding, "Instruction in Cement Laboratories," SPEE *Proc.* 20 (1912): 149–53.

49. Virgil R. Fleming, "Some Notes on Conducting Tests in Materials Testing at the University of Illinois," SPEE *Proc.* 20 (1912): 255. See also Spalding, "Instruction in Cement Laboratories," and J. Hammond Smith, "Instructive Features and Efficiency of Laboratory Courses for Undergraduate Education," SPEE *Proc.* 20 (1912): 241–45. One professor at the University of Cincinnati who disdained

preprinted laboratory forms went so far as to say that the best engineering teacher is the one "who speedily makes himself unnecessary." Robert G. Brown, "An Account of the Changes in, and Operation of, the Materials-Testing Laboratory at the University of Cincinnati," SPEE *Proc.* 20 (1912): 177. See also Francis C. Caldwell, "Laboratory Notes and Reports," SPEE *Proc.* 10 (1902): 66-72.

50. H. H. Firker, "Methods of Preparing Reports for Tests in Engineering Laboratories," SPEE *Proc.* 10 (1920): 402-3; Cowdrey and Adams, *Materials Testing Theory and Practice,* 10; Hatt, "Laboratory Course," 253-54; Herbert Gilkey, "Materials Laboratory," 1935, ISU-GP, Record Series 11/7/11, Box 2.

51. Anson Marston to President A. B. Storms, 29 January 1906, ISU-MP, Record Series 11/1/11, Box 1; Hatt, "Laboratory Course."

52. Brown, "Materials Testing Laboratory at the University of Pennsylvania," 174; George R. Chatburn, "A Laboratory Exercise: Calibration of a Riehle-Gray Apparatus," SPEE *Proc.* 8 (1901): 292-93; Gilkey et al., *Materials Testing,* 26-27.

53. Dudley, "Testing Engineer," 241. See also John Price Jackson, "Methods of Study for Technical Students," SPEE *Proc.* 11 (1903): 110-12. Jackson expressed this view in rather extreme form when he said that for engineering instruction as a whole, "informality can be good."

54. M. P. Cleghorn to Anson Marston, 5 August 1929, ISU-MP, Record Series 11/1/11, Box 1; Marston, "What Does the Engineering College Expect"; Cowdrey and Adams, *Materials Testing Theory and Practice,* iii.

55. On the origins and reception of this report, see Noble, *America by Design,* 46, 203-7.

56. Mann, "Study of Engineering Education," 14, 42-43.

57. "General Laboratory Directions for Testing Materials," Laboratory of Applied Mechanics, University of Illinois, 1904, UIA, Record Group 11/11/22.

58. W. K. Hatt, "Discussion of 'Some Observations Regarding the Methods of Teaching the Theory of Reinforced-Concrete Design,' by Professor A. B. McDaniel," *Bulletin of the Society for the Promotion of Engineering Education* 8 (1917): 409.

59. Charles A. Ellis, "Reinforced-Concrete Theory without the Aid of Formulas," *Bulletin of the Society for the Promotion of Engineering Education* 9 (1918): 381.

60. Even where the demonstration of relationships between theory and practice was given very low priority, it was still considered a significant pedagogical tool in classroom situations in which testing equipment and time for testing were limited. See Thompson, "Testing of Engineering Materials," 191.

61. Reich, *Making of American Industrial Research,* 24; Seely, "Research, Engineering, and Science," 359-62.

62. Ellis, "Reinforced-Concrete Theory," 461.

63. Ibid., 463.

64. Timoshenko, *History of Strength of Materials,* 190-97. For additional history of graphical analysis in engineering, see Edwin T. Layton Jr., "The Dimensional

Revolution: The New Relations between Theory and Experiment in the Age of Michelson," in *The Michelson Era in American Science, 1870-1930,* ed. Stanley Goldberg and Roger H. Stuewer (New York: American Institute of Physics, 1988), 23-41.

65. Ellis, "Reinforced-Concrete Theory," 466. Similarly, Jackson, "Methods of Study," 101-2.

66. Mann, "Study of Engineering Education," 63.

67. "A Practical Engineers View," *Lewis Institute Bulletin* 3 (January 1905): 2.

68. *Engineering Education* 8 (1917): 157.

69. Wickenden, "Comparative Study," 256-57.

70. See Seely, *Building the American Highway System;* Daston and Galison, "Image of Objectivity"; and T. F. Gieryn, "Boundary-Work and the Demarcation of Science from Non-science: Strains and Interests in the Professional Ideologies of Scientists," *American Sociological Review* 48 (1983): 781-93. For a broader view of the moral components of Progressive Era expertise, see Frederick C. Mosher, *Democracy and the Public Service* (New York: Oxford University Press, 1982), on the self-identification of public administrators (71-82).

71. See Robert E. Kohler, *Lords of the Fly: Drosophila Genetics and the Experimental Life* (Chicago: University of Chicago Press, 1984), for similar dynamics surrounding material and intellectual problems in biology.

72. Everett Hughes, *The Sociological Eye* (Chicago: Aldine Atherton, 1971), 379.

73. Mead, *Contracts, Specifications, and Engineering Relations,* 9.

74. Cowdrey and Adams, *Materials Testing Theory and Practice,* 3.

75. Hughes, *Sociological Eye,* 379. The success the educators had in this area is reflected in two sources. The correspondence of university engineering departments such as Penn's show numerous requests by engineering and materials firms for the names of students who might fill jobs. More broadly, biographies of members of the professional engineering societies show graduates of these departments employed in industries, frequently as directors of large construction projects. See note 32, above.

76. Reich *(Making of American Industrial Research)* and Noble *(America by Design)* both make this case. Layton describes material sciences as having been performed with less rigor than physics, and still less when instruments were involved ("Mirror-Image Twins," 575). Mowery and Rosenberg refute the idea that industrial advance was foiled by this analytical work *(Technology and the Pursuit of Economic Growth,* 35-97).

77. Mowery and Rosenberg, *Technology and the Pursuit of Economic Growth,* 35-97. Charles Rosenberg, discussing early university researches on agricultural problems, makes a similar point. He explains that "a tradition of client-centered research led to much trivial and redundant work" in university laboratories, but the agricultural experiment station "played in sum a positive role in the develop-

ment of the scientific disciplines in the United States," actually helping prepare the academic research context for more fundamental work. Rosenberg, "Science, Technology, and Economic Growth," 168.

78. Thompson, "Testing of Engineering Materials," 191.

79. J. A. L. Waddell, "Some Important Questions in Engineering Education," SPEE *Proc.* 23 (1915): 214; see also Bruce Seely, "SHOT, The History of Technology, and Engineering Education," *Technology and Culture* 39 (1995): 739-72.

80. Anson Marston, "The Engineer's Responsibility to the State," ms., 24 June 1931, ISU-MP, Record Series 11/1/11, Box 5, 3.

81. Anson Marston, "Has Specialization Been Overdone in the College Course?" undated ms., ISU-MP, Record Series 11/1/11, Box 4.

82. Paula Fass, *Outside In: Minorities and the Transformation of American Education* (New York: Oxford University Press, 1989).

83. Marston, "What Does the Engineering College Expect"; Mack, "Trades Training," 312; "Practical Engineers View," 2.

84. Jackson, "Methods of Study," 104.

85. Dudley, "Testing Engineer," 234; Anson Marston, "Engineering Education at State Technical Colleges," *Iowa Engineer* 13 (1912): 157. Charles Riborg Mann celebrated the similarities between engineering and law that might result if the former were to be taught on a "case" basis. Mann, "Study of Engineering Education," vi-vii.

86. Waddell, "Some Important Questions," 209-11.

87. "Practical Engineers View," 2.

88. "Engineering Work of the Lewis Institute," *Lewis Institute Bulletin* 1 (May 1903): 1.

89. Frank O. Marvin, "Address by the President," SPEE *Proc.* 8 (1901): 14-15. See also Mead, *Contracts, Specifications, and Engineering Relations*, 9.

90. Jackson, "Methods of Study," 102.

91. Ernest McCollough, *Transactions of the American Society of Civil Engineers*, December 1912, cited by Anson Marston in "Engineering Education at State Technical Colleges and at Institutes of Technology," *Iowa Engineer* 13 (1912): 156-59.

92. Anson Marston, talk presented to Southwest Power Conference, 7 September 1931, ISU-MP, Record Series 11/1/11, Box 9, n.p.

93. Sinclair, *Centennial History;* Noble, *America by Design;* Edwin Layton, *Revolt of the Engineers* (1971; reprinted Baltimore: Johns Hopkins University Press, 1986).

94. "Graduates of the Div. of Engineering, ISC," typescript, 1912, ISU-MP, n.p.

95. Mary Frank Fox, "Women and Scientific Careers," in *Handbook of Science and Technology Studies*, ed. Sheila Jasanoff et al. (London: Sage Publications, 1995), 205-8. See also Betty M. Vetter, "Women Scientists and Engineers: Trends in Participation," *Science* 214 (1981): 1313-21.

96. *Lewis Institute Bulletin* 2 (1904): 6.

97. Elmina Wilson to Anson Marston, 3 April 1904, ISU-MP, Record Series 11/1/11, Box 1. Wilson's family origins are suggestive. Her sister was companion to the women's rights activist Carrie Chapman Catt; Catt's husband was a prominent alumnus of Iowa's engineering school.

98. Almon H. Fuller, *A History of Civil Engineering at Iowa State College* (Ames: Iowa State University, 1959), 16.

99. Mead, *Contracts, Specifications, and Engineering Relations*, 66.

100. Anson Marston, "The Qualifications of Engineers," *Iowa Engineer* 17 (1916): frontispiece; Herbert J. Gilkey, "Staff Memo," 22 October 1935, ISU-GP, Record Series 11/7/11, Box 4; President Hall, Clark University, "Men Teachers," *Lewis Institute Bulletin* 4 (1906): 7.

101. Gilkey, "Staff Memo." Similarly, Mead, *Contracts, Specifications, and Engineering Relations*, 16.

102. Mead, *Contracts, Specifications, and Engineering Relations*, 23–26.

103. Fass, *Outside In*. See also David A. Hollinger, "Inquiry and Uplift: Late Nineteenth-Century American Academics and the Moral Efficacy of Scientific Practice," in *The Authority of Experts*, ed. Thomas L. Haskell (Bloomington: Indiana University Press, 1984), 142–56; and David E. Wharton, *A Struggle Worthy of Note: The Engineering and Technical Education of Black Americans* (Westport, Conn.: Greenwood Press, 1992).

104. Williams, *Building an Engineering Career*, 42. See also Mead, *Contracts, Specifications, and Engineering Relations*, 17, on the idea of "native ability."

105. Dudley, "Testing Engineer," 239–40.

106. Jackson, "Methods of Study," 108–9.

107. "Engineering Work of Lewis Institute," 1; Michael Katz, *The Deserving Poor* (New York: Pantheon, 1989), 9–35.

108. Charles R. Mann, "What Is an Engineer?" *Engineering Record* 74 (1917): 10.

CHAPTER TWO. Science on Site

Epigraph: "Report of Committee C-1 on Standard Specifications for Cement," ASTM *Proc.* 13 (1913): 240–41.

1. The idea that technical experts employed by industry function as "information brokers" is developed in Geoffrey Bowker, *Science on the Run: Information Management and Industrial Geophysics at Schlumberger, 1920–1940* (Cambridge, Mass.: MIT Press, 1994). This concept dovetails neatly with Steven Shapin's delineation of the role of literary technologies in scientific practice; see "Pump and Circumstance: Robert Boyle's Literary Technology," *Social Studies of Science* 14 (1984): 481–520.

2. For relevant descriptions of changing industrial workplace organization after 1900, see Harry Braverman, *Labor and Monopoly Capital: The Degradation of Work in the Twentieth Century* (New York: Monthly Review Press, 1974), particularly

chap. 4, "Scientific Management"; Richard C. Edwards, "The Social Relations of Production in the Firm and Labor Market Structure," in *Labor Market Segmentation*, ed. Richard C. Edwards, Michael Reich, and David M. Gordon (Lexington, Mass.: D. C. Heath, 1973); and Marc Silver, *Under Construction: Work and Alienation in the Building Trades* (Albany: SUNY Press, 1986).

3. On relations in the construction industries after 1900, see William Haber, *Industrial Relations in the Building Industry* (1930; reprinted New York: Arno and New York Times, 1971), especially part 2, "The Industrial Setting"; Harry C. Bates, *Bricklayers' Century of Craftsmanship* (Washington, D.C.: Bricklayers, Masons, and Plasterers International Union of America, 1955); and Gwendolyn Wright, *Moralism and the Model Home: Domestic Architecture and Cultural Conflict, 1873-1913* (Chicago: University of Chicago Press, 1980), especially chap. 6, "The Political Setting."

4. "Report of the Committee on Standard Specifications for Cement," ASTM *Proc.* 11 (1911).

5. Richard Shelton Kirby, *The Elements of Specification Writing* (New York: John Wiley and Sons, 1935); John C. Ostrup, *Standard Specifications for Structural Steel, Timber, Concrete, and Reinforced Concrete* (New York: McGraw-Hill, 1911); Daniel W. Mead, *Contracts, Specifications, and Engineering Relations* (1916; reprinted New York: McGraw-Hill, 1933).

6. David C. Mowery and Nathan Rosenberg, "The Beginnings of the Commercial Exploitation of Science by U.S. Industry," chap. 3 in *Technology and the Pursuit of Economic Growth* (Cambridge: Cambridge University Press, 1989); David Noble, *America by Design: Science, Technology, and the Rise of Corporate Capitalism* (New York: Oxford University Press, 1977), 121-28; Bruce Sinclair, *Early Research at the Franklin Institute: The Investigation into Causes of Steam Boiler Explosions, 1830-1837* (Philadelphia: Franklin Institute, 1966); David Hounshell, *From the American System to Mass Production, 1800 to 1932* (Baltimore: Johns Hopkins University Press, 1984), 93-99, 106.

7. See "The American Society for Testing Materials: Its Purpose and Work," *Index to ASTM Standards and Tentative Specifications* (Philadelphia: ASTM, 1935), 5.

8. C. L. Warwick, "The Work in the Field of Standardization of the American Society for Testing," *Annals of the American Academy of Political and Social Science* 37 (1928): 49-54.

9. Rexmond C. Cochrane, *Measures for Progress: A History of the National Bureau of Standards* (New York: Arno Press, 1976), 14-18, 253-63; Samuel Krislov, *How Nations Choose Product Standards and Standards Change Nations* (Pittsburgh: University of Pittsburgh Press, 1997), 26-52; Gustavus Weber, *The Bureau of Standards: Its History, Activities, and Organization* (Baltimore: Johns Hopkins Press, 1925), 44-55; Noble, *America by Design*, 71-76; A. Hunter Dupree, *Science in the Federal Government* (Cambridge, Mass.: Harvard University Press, 1957), 272.

10. Krislov, *How Nations Choose*, 27-37.

11. The U.S. Army Corps of Engineers participated in the general exchange of technical knowledge about materials by publishing its own specifications in trade journals, and by joining the American Concrete Institute and similar bodies, but was not otherwise an arbiter of building or manufacturing practice. See William L. Marshall et al., "Testing Hydraulic Cements—II," *Engineering News* 44 (21 September 1901): 275–77.

12. Mowery and Rosenberg, *Technology and the Pursuit of Economic Growth*, 46–47; "Basis of Cooperation between Various Government Branches and the Standing Committees of the Society," ASTM *Yearbook* (Philadelphia: ASTM, 1922), 245.

13. Warwick, "Work in the Field of Standardization," 50–51.

14. Ibid., 50–54; Daniel J. Hauer, "Specifications," in *Handbook of Building Construction*, vol. 2, ed. George A. Hool and Nathan C. Johnson (New York: McGraw-Hill, 1920), 1075–76. New York City included ASTM standards for steel in its solicitation of bids for the new subway in 1902; a small construction company incorporated them into a contract written for a private client in the same year. "The New York Rapid Transit Railway, XI: Inspection of Cement," *Engineering News* 48 (25 September 1902): 242–43; contracts of the Foundation Company, Chicago, Foundation Company records, Box 9, Division of Engineering and Industry, National Museum of American History, Washington, D.C. Not everyone was pleased with the spreading practice, however. On the proliferation of ASTM specifications and mixed reactions to their use by some building contractors, see Leonard C. Wason, "The Problems of the Contractor," ACI *Proc.* 11 (1914): 372. Wason found materials specifications to be generally too rigorous and unsuited to conditions on the construction site. See also Krislov, *How Nations Choose*, 3–79, and Sara Wermiel, *The Fireproof Building: Technology and Public Safety in the Nineteenth-Century American City* (Baltimore: Johns Hopkins University Press, 2000).

15. William B. Coney and Barbara Posadas, "Concrete in Illinois: Its History and Preservation," Illinois Preservation Series of the Illinois Historic Preservation Agency, no. 8 (n.d.).

16. Jasper O. Draffin, "A Brief History of Lime, Cement, Concrete, and Reinforced Concrete," *Journal of the Western Society of Engineers* 48 (1943), reprinted in *A Selection of Historic American Papers on Concrete, 1876–1926*, ed. Howard Newlon Jr. (Detroit: American Concrete Institute, 1976), 27. Duff Abrams, working for the Portland Cement Association in 1918, concluded from some fifty thousand tests that "for the given materials the strength depends upon only one factor—the ratio of cement to water." Duff Abrams, "Design of Concrete Mixes," Minutes of the Association of Manufacturers of Portland Cement, December 1918, reprinted in ibid., 306–10. Other researchers contested this, ultimately concluding that the ratio of cement to sand and stone figured in the strength of a concrete mixture.

17. C. B. Porter, "Railway Concrete," *JACI* 23 (May 1952): 726. Even though

dryer mixtures proved stronger in laboratory tests, in practical conditions the ease of placement afforded by wetter concretes made them more reliable.

18. Halbert P. Gillette and George S. Hill, *Concrete Construction Methods and Cost* (New York: Myron C. Clark, 1908), 25–27.

19. For a typical case of routinization in the concrete construction industry, see promotional materials of the Trussed Concrete Steel Company, including descriptions of prefabricated steel reinforcing, in "Hy-Rib: Its Application in Roofs, Floors, Wall, Sidings, Partitions, Ceilings, Furring" (Detroit, 1912), Trade Catalog Collection of Hagley Museum and Library, Wilmington, Del. The use of prefabricated reinforcement would bring great economies to builders and building buyers, the company claimed, by permitting the dismissal of highly skilled steel handlers. Also Henry Haag, "Field Notes of a Concrete Gang," *Iowa Engineer* 7 (1907): 69–73.

20. Alfred Chandler and other analysts of modern management except construction from the early-twentieth-century industrial embrace of mass-production methods. Alfred Chandler, *The Visible Hand: The Managerial Revolution in American Business* (Cambridge, Mass.: Harvard University Press, 1977), 242; Arthur Stinchcombe, "Bureaucratic and Craft Administration of Production: A Comparative Study," *Administrative Science Quarterly* 4 (1959): 169. Their view is contradicted not only by large-scale concrete construction projects, such as the Panama Canal and Gropius's massive Dessau-Torten housing project of the late 1920s, but by the countless smaller commercial projects that used routinized labor and standardized or prefabricated building parts after 1900. See Silver, *Under Construction*, 317–19.

21. Mario Salvadori, *Why Buildings Stand Up: The Strength of Architecture* (New York: Norton, 1980), 67.

22. William A. Maples and Robert E. Wilde, "A Story of Progress: Fifty Years of the American Concrete Institute," *JACI* 25 (February 1954): 412–13. A 1904 editorial in *Cement Age,* a journal produced by and for the cement industry, praised contemporary German practices of strict supervision and hiring for the building trades: "Steps of this kind are certainly steps in the right direction, for most failures in concrete work are not likely to result from lack of strength in the concrete itself, nor from want of proper calculations in the Engineering Department, but from lack of supervision and watchfulness on the part of those charged with the examination of the work." *Cement Age* 1 (December 1904): 228.

23. Haber explains that through the 1920s, employers in the building industry were more concerned with speed than with having thoroughly trained workers. He addresses specialization among unions as a product of this concern for speed and shows how it leads to rigid jurisdictional definitions and the dispossession of building workers by "helpers" and "greenhands." Haber, *Industrial Relations and the Building Industry,* 47.

24. See ibid., chap. 3, "Competition for Jobs—The Building Business." In fact, contractors and local industrialists *were* willing to train workers; they sponsored building trades schools in many cities, but these represented vocational programs that deemphasized experimentation and decision making. Wright, *Moralism and the Model Home*, 175. By the 1920s Iowa State College was offering courses for the training of foremen, but the engineering department stated that "the purpose of the courses is to teach the foreman his responsibility as manager, supervisor and instructor rather than to teach him the technology of manufacturing his particular product." Technical knowledge remained a separate stratum of skill. "Foreman Training in Industrial Establishments," Iowa State College of Agriculture and Mechanical Arts Official Publication 25 (1927), n.p.

25. Mead, *Contracts, Specifications, and Engineering Relations*, 314

26. For a thorough description of one such laboratory, see "New York Rapid Transit Railway, XI."

27. Some consumers of cement tested their purchases at the cement mill, prior to shipping. This speeded up delivery and prevented large quantities of cement from taking up storage space at the construction site. See Richard K. Meade, *Manufacture of Cement* (Scranton, Pa.: International Textbook Co., 1922), part 2.

28. "Simple and Rapid Tests for Cement," *Engineering News* 48 (4 September 1902): 166, describing tests that can be made "with an outfit costing not over $4, and which can be stored in a desk pigeonhole."

29. Ernest McCullough, "Proper Specifications and Design Standards for Reinforced Concrete," *Twenty-eighth Annual Report of the Illinois Society of Engineers and Surveyors,* 1913, 205; Norman N. Aylon, "Concrete Construction Inspection: Who Does What?" *Canadian Architect* 19 (July 1974): 38. A word about failed tests: Many tests, by their nature, yielded results only after a cement or a concrete mix had been incorporated into the building under construction. Unsatisfactory results might occasionally result in the replacement of an element or in the redesign of a building to add buttressing of some type, but most often they brought an adjustment of fees or other compensation by the contractor. Some specifications included provisions for arbitration between contractor and building buyer on such matters. See C. K. Smoley, *Stone, Brick, and Concrete* (Scranton, Pa.: International Textbook Co., 1928), 14, 38.

30. H. C. Berry to A. Noble and Son, 19 December 1910, and A. Noble to H. C. Berry, 21 December 1910, General Correspondence File, Towne Scientific School, UPA, Record Group UPD 2.2, Box 1.

31. McCullough, "Proper Specifications and Design Standards," 205. A 1913 catalog of concrete construction machinery offered two "field models" of compression-testing machines. One could test at forces of 50 tons, the other at 75 tons. Catalog of the F. H. Hopkins Company, 1913, ACC Archives.

32. "Report of the Board of Engineers, U.S.A., on the Properties and Testing of Hydraulic Cement," *Engineering News* 46 (12 September 1901): 181.

33. Pitson Cleaver, in *Engineering News* 48 (28 October 1902): 382; Robert Lesley, "Ninth Annual Convention of the ASTM," *Cement Age* 13 (July 1906): 110.

34. Described in H. F. Gonnerman, "Development of Cement Performance Tests and Requirements," *Research Department Bulletin, Portland Cement Association* 3 (March 1958): 18.

35. Edouard Candlot, "Tests of Materials of Construction: Materials Other Than Metals," *Cement Age* 1 (December 1904): 236–37.

36. "The Proper Manipulation of Tests of Cements: A Symposium," *Engineering News* 43 (21 June 1900): 408–9.

37. Ibid.

38. "Progress Report of Special Committee on Uniform Tests of Cement," *Proceedings of the American Society of Civil Engineers* 29 (January 1903): 2–11.

39. "American Society for Testing Materials: Its Purpose and Work," 5.

40. "Report of the Committee on Standard Specifications for Cement," ASTM *Proc.* 5 (1905): 77.

41. John G. Brown, "Discussion of Specifications for Cement," ASTM *Proc.* 5 (1905): 124; Robert W. Lesley, "Discussion on Standard Specifications for Cement," ASTM *Proc.* 7 (1907): 132.

42. ASTM *Proc.* 17 (1917): 513.

43. As Stuart Shapiro has pointed out, a number of historians of engineering have noted that within standards there can reside possibilities for interpretation, but all treat this as an issue of creativity—a means of allowing design freedom, or "room to maneuver," within an organized system of economic relations. Social contestation among occupations and classes does not enter into these analyses. Stuart Shapiro, "Degrees of Freedom: The Interaction of Standards of Practice and Engineering Judgment," *Science, Technology, and Human Values* 22 (1997): 291–93, discussing William Addis, *Structural Engineering: The Nature of Theory and Design* (London: Ellis Horwood, 1990), and Walter Vincenti, *What Engineers Know and How They Know It* (Baltimore: Johns Hopkins University Press, 1990). Andrew Abbott recognizes the occupational function of standardization in delineating a jurisdiction that is neither too broad nor too narrow to be realistically controlled, but he treats this task as a reaction primarily to vertical competition perceived within an occupation rather than to competition felt between occupations. Andrew Abbott, *The System of Professions: An Essay on the Division of Expert Labor* (Chicago: University of Chicago Press, 1988), 51–52.

44. "Standard Methods of Testing Cement," *Engineering Record* 34 (24 October 1896).

45. For other sociologically informed discussions of scientific testing, see Trevor Pinch, "'Testing—One, Two, Three . . . Testing': Toward a Sociology of

Testing," *Science, Technology, and Human Values* 18 (1993): 25–41, and Donald Mackenzie, "The Construction of Technical Facts," chap. 7 in *Inventing Accuracy: An Historical Sociology of Nuclear Missile Guidance* (Cambridge, Mass.: MIT Press, 1990).

46. "Report of the Board of Engineers, U.S.A.," 181.

47. See, for example, Richard L. Humphrey, comments in "Discussion on Specifications for Cement," ASTM *Proc.* 4 (1904): 135.

48. W. Purves Taylor, "Notes on Cement Testing," *Papers Read before the American Association of Portland Cement Manufacturers* (New York: Portland Cement Association, 1904), 36. On the history of the "personal equation," see Simon Schaffer, "Astronomers Mark Time: Discipline and the Personal Equation," *Science in Context* 2 (1988): 115–45. Frederick W. Taylor supported the quantification of such personal idiosyncrasy through the determination of "personal coefficients," a development of physiologists that measured reaction times in production workers. Frederick W. Taylor, *The Principles of Scientific Management* (1911; reprinted New York: Norton, 1967), 89–94.

49. Richard L. Humphrey, "Cement Testing in Municipal Laboratories," ASTM *Proc.* 2 (1902): 156.

50. Mead, *Contracts, Specifications, and Engineering Relations*, 297.

51. Ibid., 90.

52. Ibid., 298.

53. Herbert Gilkey, Glenn Murphy, and Elmer O. Bergman, *Materials Testing* (New York: McGraw-Hill, 1941), v.

54. Ibid., 104.

55. Irving Cowdrey and Ralph Adams, *Materials Testing Theory and Practice* (New York: John Wiley and Sons, 1925), 4. See also Mead, *Contracts, Specifications, and Engineering Relations*, 3, 31–34, 295.

56. Taylor, "Notes on Cement Testing," 36.

57. Humphrey, "Cement Testing in Municipal Laboratories," 135.

58. Paul Kreuzpointner, "The Ethics of Testing," ASTM *Proc.* 2 (1902): 119.

59. "Progress Report of Special Committee on Uniform Tests of Cement."

60. ASTM *Proc.* 4 (1904): 114, 118.

61. Many examples of this ideology can be found in the communications of engineering educators in this period. References to the importance of inducing "surprise" or "shock"—by allowing the students to discover errors in instrument calibration or by having them witness the dramatic destruction of a test specimen—are especially suggestive. George R. Chatburn, "A Laboratory Exercise: Calibration of a Riehle-Gray Apparatus," SPEE *Proc.* 8 (1901), 292–93; Robert G. Brown, "An Account of Change in, and Operation of, the Materials Testing Laboratory of the University of Cincinnati," *Engineering Education* 20 (1912): 174.

62. "Standard Specifications for Cement," ASTM *Yearbook* (Philadelphia: ASTM, 1911), 140.

63. Ernest McCready, "Some Avoidable Causes of Variation in Cement Testing," ASTM *Proc.* 7 (1907): 351–52.

64. Ibid., 349.

65. ASTM *Yearbook*, 1911, 137.

66. Marshall et al., "Testing Hydraulic Cements—II," 250.

67. Allen W. Carpenter, "A Ramming Device for Compacting Cement in Briquette Molds," *Engineering News* 47 (9 January 1902): 30; C. J. Griesenaur, "A Laboratory Mixer for Dry Cement and Sand," *Engineering News* 48 (7 August 1902): 100.

68. ASTM, "Standard Specifications for Cement, Serial Designation: C1-09" (adopted 1904; revised 1908, 1909), 303.

69. Haber, *Industrial Relations and the Building Industry*, 33. Concrete had been mixed mechanically since the 1890s, but in certain contexts hand-mixing was used into the 1930s. It persisted where very small amounts of concrete were needed or where a site was too difficult to traverse with heavy machines. See George Underwood, *Standard Construction Methods* (New York: McGraw-Hill, 1931), 127–29, and Gillette and Hill, *Concrete Construction*, 43, 61. The logistics of concrete handling are further discussed in chapter 4, below.

70. John C. Wait, cited in Mead, *Contracts, Specifications, and Engineering Relations*, 96.

71. This "visibility" is associated with accomplishment and contrasts with the negative visibility granted to erring technicians within laboratory settings. I draw here on Steven Shapin, "The Invisible Technician," *American Scientist* 77 (1989): 554–62.

72. *Engineering News Supplement* 50 (1904): 269, cited in Mead, *Contracts, Specifications, and Engineering Relations*, 284.

73. Gillette and Hill, *Concrete Construction*, 29, on tests performed by the state engineer of Illinois.

74. Mead, *Contracts, Specifications, and Engineering Relations*, 281.

75. David A. Hollinger, "Inquiry and Uplift: Late Nineteenth-Century American Academics and the Moral Efficacy of Scientific Practice," in *The Authority of Experts*, ed. Thomas L. Haskell (Bloomington: Indiana University Press, 1984), 142–56.

CHAPTER THREE. Science and the "Fair Deal"

Epigraph: Daniel W. Mead, *Contracts, Specifications, and Engineering Relations* (1916; reprinted New York: McGraw-Hill, 1933), 3.

1. Frank Kerekes and Harold B. Reid Jr., "Fifty Years of Development in Building Code Requirements for Reinforced Concrete," *JACI* 25 (February 1954): 442.

2. "Report of Committee C-1 on Standard Specifications for Cement," ASTM *Proc.* 13 (1913): 231. This USGS cooperation was attributed by the ASTM to the fact that Richard Humphrey was both the secretary of the joint committee and the director of the St. Louis laboratory.

3. Ibid., 230.

4. Kerekes and Reid, "Fifty Years of Development," 442–43; "Final Report of the Special Committee on Concrete and Reinforced Concrete," *Transactions of the ASCE* 81 (1917): 1106–82.

5. "Report of Committee C-1," 230–39.

6. Cecil D. Elliott, *Technics and Architecture* (Cambridge, Mass.: MIT Press, 1992), 184.

7. H. F. Gonnerman, "Development of Cement Performance Tests and Requirements," *Portland Cement Association Bulletin* 93 (March 1958): 23.

8. *Cement Age* 2 (July 1908): 11. The reported survey, distributed by the ASTM's Committee for Standard Specifications on Cement, found that of 143 engineers, architects, and contractors, 93 used the ASTM's specifications, 30 used specifications of their own devising, and 12—all of whom were army officers in charge of cement work—used specifications of the U.S. Army Corps of Engineers; 8 did not use cement in their work.

9. For example, see "Predicting Concrete Strengths" and "Testing of Concrete Materials and Concrete," unpublished instruction booklets, 1928, of the Foundation Company, Chicago, Foundation Company records, Box 9, Division of Engineering and Industry, National Museum of American History, Washington, D.C.

10. "Annual Convention of the American Society for Testing Materials," *Cement Age* 3 (July 1906): 109; C. L. Warwick, "The Work in the Field of Standardization of the American Society for Testing Materials," *Annals of the American Academy of Political and Social Science* 37 (1928): 53.

11. ASTM, *Index to ASTM Standards and Tentative Standards* (Philadelphia: ASTM, 1935), 5.

12. Warwick, "Work in the Field of Standardization," 58–59.

13. ASTM *Yearbook* (Philadelphia: ASTM, 1914), 8.

14. Bruce Sinclair makes this point about the ASME in *A Centennial History of the American Society of Mechanical Engineers, 1880–1980* (Toronto: ASME, 1980), 153. See also David Noble, *America by Design: Science, Technology, and the Rise of Corporate Capitalism* (New York: Oxford University Press, 1977), 81–82. Edwin Layton contrasts the identity of the ASME between 1900 and 1910 as an "agent of industry" with that of the American Institute of Electrical Engineers, which barred businessmen from full memberships at the time. The AIEE maintained an emphasis on professional unity rather than industrial accommodation until pressure from industry precipitated a crisis in 1912 and 1913. Edwin Layton, *The Revolt of the Engineers* (Baltimore: Johns Hopkins Press, 1971), 37–40.

15. NBS *AR*, 1915, 117–19; NBS *AR*, 1920, 191. S. W. Stratton, the director of the NBS from its founding through 1923, directed his staff's attention increasingly towards studies of cement. The bureau's work included field and laboratory tests on the permeability of cement in water; tests on the effects of heat, moisture, and pressure on cement; and assessments of the specific gravity and fineness of cement samples (these were thought to indicate the purity and strength of a cement). Using the latest machinery, bureau engineers tested cements' setting times and performed some of the first tests of "live loads" on concrete floors. The scope of the bureau's ambitions and resources is indicated by the fact that in 1916 the bureau began producing its own cements for testing, creating seventy-four varieties in a Minnesota plant. Cement laboratories of the bureau and the USGS were combined in 1902. NBS *AR*, 1912, 25–26; NBS *AR*, 1914, 75; NBS *AR*, 1916, 130; Henry J. Cowan, *The Master Builders* vol. 1, *Science and Building: Structural and Environmental Design in the Nineteenth and Twentieth Centuries* (New York: John Wiley and Sons, 1977), 94.

16. Noble identifies a reforming spirit in the engineers who worked on standardization through the ASME and ASTM; they "were best able to transcend the narrow self-interest of any specific company and view the industry as a whole." Noble, *America by Design,* 77. Sinclair attributes the same attitude to members of the ASME. Sinclair, *Centennial History,* 55.

17. Robert W. Lesley, "Discussion on Standard Specifications for Cement," ASTM *Proc.* 7 (1907): 132.

18. Sinclair describes the self-consciousness of professionals in the ASME as they developed codes for boiler safety and other technologies: "These men certainly knew about the economic power of specialized knowledge." Sinclair, *Centennial History,* 153.

19. *Cement Age* 1 (March 1905): 453.

20. Portland Cement Association, *Portland Cement Association Reference Book for Editors 1927* (Chicago: Portland Cement Association, 1927), 2–6.

21. Portland Cement Association, "When Science Goes to Work for Modern Industry," brochure, ca. 1924, 5–9; Portland Cement Association, *Portland Cement Association Reference Book for Editors 1925* (Chicago: Portland Cement Association, 1925), 16–17. The PCA's research facilities at the Lewis Institute, called the Structural Materials Research Laboratory, were directed by Professor Duff Abrams of the University of Illinois. In addition to its chemical and structural testing, the laboratory maintained a much publicized "sand library" of three thousand samples of sands for use in cement mixes. Such samples allowed cement manufacturers from different parts of the United States and abroad to learn the quality of sand nearest their plant. Information on the quality of local raw materials was probably also of interest to those seeking to invest in cement plants, in which a boom occurred in the 1910s and 1920s. For information on the promotional

aspects of this boom, see the trade catalog collection of the Avery Architectural Library, Columbia University.

22. Portland Cement Association, *Editor's Reference Book,* 1925, 15–16.

23. Portland Cement Association, "When Science Goes to Work" and "High Lights on the Portland Cement Industry," brochure, 1924.

24. Richard Lewis Humphrey, "Results of Tests Made in the Collective Portland Cement Exhibit and Model Testing Laboratory of the Association of American Portland Cement Manufacturers" (St. Louis: Association of American Portland Cement Manufacturers, 1904), 1–10.

25. Richard L. Humphrey and C. F. McKenna, "Discussion on Standard Specifications," ASTM *Proc.* 7 (1907): 134–35.

26. George S. Webster, "Discussion on Standards Specifications," ASTM *Proc.* 7 (1907): 137.

27. Herbert Gilkey, *Anson Marston: ISU'S First Dean of Engineering* (Ames: Iowa State University, College of Engineering, 1968), 5.

28. Anson Marston, "The Engineer's Responsibility to the State," talk presented to the American Society of Automotive Engineers, 24 June 1931, ISU-MP, Record Series 11/1/11, Box 5, 3.

29. Arthur N. Talbot, "The Extension of Engineering Investigational Work by Engineering Schools," SPEE *Proc.* 12 (1904): 81.

30. Charles E. Rosenberg, "Science, Technology, and Economic Growth: The Case of the Agricultural Experiment Station Scientist, 1875–1914," in *No Other Gods* (Baltimore: Johns Hopkins Press, 1961), 154.

31. Aluminate Patents Company to H. C. Berry, 15 September 1911, and reply, 16 September 1911, UPA, Record Group UPD 2.1, Box 2, File: "Engineering Correspondence July 1910–September 1911."

32. H. C. Berry to David Ash, 24 April 1911, UPA, Record Group UPD 2.1, Box 2, File: "Engineering Correspondence July 1910–September 1911."

33. For example, F. H. Souder to H. C. Berry, 14 June 1910, regarding specimens sent from Spackman Engineering Company to Penn's laboratory, UPA, Record Group UPD 2.2, Box 1, File: "Engineering Correspondence June 1909–January 1912." To some degree state university laboratories had to be more conscious of their relations with local testing businesses than did laboratories of private schools. The public institutions were charged with using public money for the benefit of state industries and could receive complaints from legislators if their activities hindered competition from testing firms.

34. Richard Perry Spence to Hon. J. B. Hungerford, Board of Trustees, Iowa State College, 9 June 1903, ISU-MP, Record Series 11/1/11, Box 1. Intriguingly, Marston seems to have claimed that his septic systems used a process that "was a natural one and therefore unpatentable," a formulation that Cameron's lawyers labeled "lawless." Anson Marston, "Outline of Steps to Be Taken in Securing

Patents on the Investigations in Connection with the College Research Work," memo to engineering faculty, 19 November 1925, ISU-MP, Record Series 11/1/11, Box 1.

35. H. C. Berry to A. Wissler Instrument Works, St. Louis, 4 October 1910, and H. C. Berry to A. Nacke and Son, Philadelphia, 19 December 1910, UPA, Record Group UPD 2.2, Box 1, File: "Engineering Correspondence June 1909–January 1912"; Lt. Col. C. B. Wheeler, Ordnance Department, Watertown Arsenal, to H. C. Berry, 22 December 1910, same file; Irving Cowdrey and Ralph Adams, *Materials Testing Theory and Practice* (New York: John Wiley and Sons, 1925), 94–95; Herbert Gilkey, Glenn Murphy, and Elmer O. Bergman, *Materials Testing* (New York: McGraw-Hill, 1941), 17.

36. Berry used a form letter to inform several university and government laboratories of his recommendations for testing equipment. That letter included recommendation of Berry's own extensometer for long-term tests of concrete beams and columns. See UPA, Record Group UPD 2.2, Box 1, File: "Engineering Correspondence June 1909–January 1912."

37. This declaration appears in correspondence of Marburg and Berry. Typical examples are H. C. Berry to David Ash, 24 April 1911, and H. C. Berry to J. N. Adams, Manager, Flexol Company, 17 January 1911, UPA, Record Group UPD 2.2, Box 1, File: "General Correspondence January 1909–January 1912."

38. F. H. Souder, Honest Reinforced-Concrete Culvert Company, to H. C. Berry, 14 June 1910, UPA, Record Group UPD 2.2, Box 1, File: "Engineering Correspondence June 1909–January 1912."

39. "Tests of Large Columns," press release from civil engineering department on tests of 14-foot columns built by Turner-Forman Concrete Steel Company, Philadelphia, UPA, Record Group UPD 2.2, Box 1, File: "General Correspondence January 1909–January 1912"; "Chicago Building Tests" (advertisement), *Cement Era*, 1924, 14.

40. Bruce Seely has noted the pride with which the University of Illinois regarded "the unprejudiced character" of its engineering bulletins. Bruce Seely, "Research, Engineering, and Science in American Engineering Colleges, 1900–1960," *Technology and Culture* 34 (1993): 350. For more on Progressive Era faith in technical expertise, see Bruce Seely, *Building the American Highway System* (Philadelphia: Temple University Press, 1987), and Samuel P. Hays, *Conservation and the Gospel of Efficiency: The Progressive Conservation Movement, 1890–1920* (Cambridge, Mass.: Harvard University Press, 1959).

41. E. Marburg to F. G. Cunningham, Fort Worth and Denver City Railway Company, 25 August 1910, and E. Marburg to J. S. Francis, Bell Telephone Company, 19 March 1910, UPA, Record Group UPD 2.2, Box 1, File: "General Correspondence January 1909–January 1912."

42. G. Saxton Thompson, "Testing of Engineering Materials," SPEE *Proc.* 20

(1912): 202. See also Anson Marston, "Tests of Sand Lime and Concrete Lime Bricks and Concrete Building Blocks," *Iowa Engineer* 4 (1904): 108–10.

43. E. R. Schreiter, City of Detroit, Common Council Committee, to E. Marburg, 17 September 1910, and reply, 20 September 1910, UPA, Record Group UPD 2.2, Box 1, File: "General Correspondence January 1909–January 1912."

44. Charles B. Dudley, "The Testing Engineer," SPEE *Proc.* 13 (1906): 250.

45. Anson Marston, "The Engineering Experiment Station of Iowa State College," undated ms., ISU-MP, Record Series 11/1/11, Box 4, 4.

46. President E. W. Stanton to Anson Marston, 28 February 1912, ISU-MP, Record Series 11/1/11, Box 1.

47. "The University in Industry," *Scientific American* 122 (27 March 1920): 328.

48. See Seely, *Building the American Highway System,* 100–31.

49. *Cement Age* 1 (August 1904): 2.

50. "*Cement Age* Buys up and Consolidates with *Concrete Engineering*," *Cement Age* 12 (January 1911): 4. In 1912 *Cement Age* absorbed another competitor, *Concrete*. Other specialized journals for the concrete trades included *Rock Products* (founded 1902) and *Expanded Metal* (1905). For additional, sometimes critical treatment of concrete technologies, one could turn to the *Brickbuilder, Iron Age,* or *Stone*. The *Contractor* often summarized financial aspects of building with concrete.

51. An important adjunct to these trade publications was the great bulk of material published by the safety and insurance industries. As early as 1901, organizations such as the American Society of Safety Engineers and the Insurance Engineering Experiment Station of Boston began issuing reports on concrete, in the form of journals such as *Insurance Engineering* or research bulletins. In addition to generating their own findings, fire protection and insurance agencies disseminated the work of other institutions. The Insurance Library Association, also in Boston, amassed a huge collection of printed information on concrete and fire protection, including publications by the engineering faculty of the University of Pennsylvania. This information was useful to insurers, but it also offered cement producers and structural engineers a source for information on concrete and on the legal demands they might face in manufacturing or using concrete.

52. Allen Brett to H. C. Berry, 31 December 1910, UPA, Record Group UPD 2.2, Box 1, File: "General Correspondence January 1909–January 1912."

53. Robert W. Lesley, in *Cement Age* 1 (February 1905): 451.

54. Allen Brett to H. C. Berry, 9 April 1910, UPA, Record Group UPD 2.2, Box 1, File: "General Correspondence January 1909–January 1912."

55. Harwood Frost to E. Marburg, 12 December 1910, UPA, Record Group UPD 2.2, Box 1, File: "E. Marburg Personal Correspondence July 1910–July 1911."

56. *Engineering News* 44 (8 November 1900): 315; "Organs Are Not Technical Journals," *Cement Age* 12 (December 1911): 234.

57. Amy Slaton and Janet Abbate, "The Hidden Lives of Standards," in *Tech-*

nologies of Power, ed. Gabrielle Hecht and Michael Allen (Cambridge, Mass.: MIT Press, forthcoming).

58. Warwick, "Work in the Field of Standardization," 60.

59. Henry S. Spackman and Robert W. Lesley, "Sands: Their Relationship to Mortar and Concrete," *Cement Age* 2 (July 1908): 38.

60. "Government Standard Specifications for Portland Cement," *Cement Age* 14 (June 1912): 284.

61. "Or Equal Thereto . . . ," *Cement Age* 12 (March 1911): 126.

62. Mead, *Contracts, Specifications, and Engineering Relations,* 302.

63. "Or Equal Thereto . . . ," 126-27.

64. C. K. Smoley, "Agreements and Specifications," in *Stone, Brick, and Concrete* (Scranton, Pa.: International Textbook Co., 1928), 10.

65. Richard C. Edwards, "The Social Relations of Production in the Firm and Labor Market Structure," in *Labor Market Segmentation,* ed. Richard C. Edwards, Michael Reich, and David M. Gordon (Lexington, Mass.: D. C. Heath, 1973), 4-5. See also Olivier Zunz, *Making America Corporate, 1870-1920* (Chicago: University of Chicago Press, 1990), and JoAnne Yates, *Control through Communication: The Rise of System in American Management* (Baltimore: Johns Hopkins University Press, 1989).

66. Elliott, *Technics and Architecture,* 364-66.

67. Ibid., 364-83. Fire-extinguishing technologies were also refined. Sprinklers and other extinguishing systems proliferated after the 1870s. See Sara Wermiel, *The Fireproof Building: Technology and Public Safety in the Nineteenth-Century American City* (Baltimore: Johns Hopkins University Press, 2000).

68. Gwendolyn Wright, *Moralism and the Model Home: Domestic Architecture and Cultural Conflict, 1873-1913* (Chicago: University of Chicago Press, 1980), 223.

69. Cowan, *Master Builders,* 199-200.

70. According to one promotional brochure, "All combustible material inside the Pacific Coast Borax refinery at Bayonne, New Jersey was burned. The fire was hot enough to fuse cast iron. . . . A little plastering repaired all damage to the concrete." "Reinforced Concrete as Used in Manufacturing Plants," brochure of the Aberthaw Construction Company, Boston, 1902, ACC Archives, 31.

71. Cowan, *Master Builders,* 90.

72. "Factors of Safety," *Engineering News* 56 (6 September 1906): 258, cited in Elliott, *Technics and Architecture,* 367.

73. Cowan, *Master Builders,* 91-92.

74. Frank A. Randall Jr., "Historical Notes on Structural Safety," *JACI* 70 (October 1973): 676. Manhattan issued regulations for reinforced concrete in 1903, but the first reinforced-concrete buildings erected there are thought to have been a factory and a school built in 1905. The fact that the city predicted future use of the material suggests the generally good reputation of reinforced concrete in

1903. See Peter Collins, *Concrete: The Vision of a New Architecture* (New York: Horizon Press, 1959), 88.

75. Kerekes and Reid, "Fifty Years of Development," 444.

76. Cowan, *Master Builders,* 91; Kerekes and Reid, "Fifty Years of Development," 454. A specific example of Chicago's influence on smaller cities is the method of constructing columns with hooped reinforcement, which emerged in the Midwest around 1915. Ibid., 449.

77. "Uniform Municipal Building Laws," *Cement Age* 1 (November 1904): 210. Cities did not necessarily rush into the adoption of building codes just because their neighbors did. Youngstown, Ohio, had no city regulations on reinforced-concrete construction until a major collapse there in 1915. *Engineering News* 73 (4 February 1915): 218.

78. "Reinforced Concrete," *Cement Age* 1 (November 1904): 207.

79. Jerome Cochrane, "A Comparison of Seven Building Code Requirements for the Design of Reinforced Concrete," *Cement Age* 14 (June 1912): 298. Similarly, Edward Godfrey, "New York and Pittsburgh Building Codes," *Concrete and Cement Age* 1 (July 1912): 44.

80. R. P. Miller, "Legislation Concerning the Use of Cement in New York City," *Proceedings of the National Association of Cement Users* 2 (1906): 192.

81. Kerekes and Reid, "Fifty Years of Development," 457.

82. Records on the case of Peter C. McArdle vs. City of Chicago et al. include Abstract of Record, Appellate Court of Illinois First District, March Term, 1910; Reply Brief of Plaintiff in Error, Appellate Court of Illinois, First District, October Term, 1918; Writ of Mandamus, Circuit Court of Cook County, 1921; Abstract of Record, Appellate Court of Illinois, First District, October Term, 1934; Petition for Rehearing, Appellate Court of Illinois, First District, October Term, 1934; and Brief and Argument for Appellant, Appellate Court of Illinois, First District, October Term, 1934. See also "City Coal and City Graft, Politics and Sudden Death," *Inter Ocean,* 19 January 1910, 1 (with thanks to the McArdle family).

83. "The New York Building Code," *Cement Age* 12 (June 1911): 289; John Preston Comer, *New York City Building Control, 1800-1941* (New York: Columbia University Press, 1942), 51-54. I thank the McArdle family for these references.

84. George S. Webster, "The Advantage of Uniformity in Specifications for Cements and Methods of Testing," ASTM *Proc.* 2 (1902): 131.

85. K. J. Lundelius, Wm. Steele and Sons, to H. C. Berry, 7 October 1913, UPA, Record Group UPD 2.2, Box 1, File: "Engineering Correspondence June 1909-January 1912"; *Brickbuilder,* October 1904; *Cement Age* 1 (December 1904): 286.

86. "Code of Principles of Professional Conduct of the American Institute of Electrical Engineers," 8 March 1912, cited in Mead, *Contracts, Specifications, and Engineering Relations,* 69.

87. William A. Maples and Robert E. Wilde, "A Story of Progress: Fifty Years of the American Concrete Institute," *JACI* 25 (February 1954): 410.

88. Ibid., 411–12.

89. Kerekes and Reid, "Fifty Years of Development," 447.

90. R. S. Greenman, "Some Problems of the Cement Inspecting Laboratory," ASTM *Proc.* 7 (1907): 359.

91. McCready, "Some Avoidable Causes of Variation," 351.

92. "A Bonus System for the Purchase of Portland Cement," *Engineering News* 70 (1913): 281; Maj. William L. Marshall et al., "Testing Hydraulic Cements—I," *Engineering Record* 44 (14 September 1901): 249. Similarly, W. R. King, U.S. Army Corps of Engineers, "The Need for More Uniform Methods of Cement Testing," *Engineering Record* 34 (21 November 1896): 465.

93. Mead, *Contracts, Specifications, and Engineering Relations*, 68–69, on the AIA and the American Institute of Consulting Engineers.

CHAPTER FOUR. The Business of Building

Epigraph: "The Why and How of a Great Plant: The Bay State Cotton Corporation Lowell Division," Aberthaw Construction Company brochure, 1920, 12.

1. For examples of this idea, see Willard L. Case, *The Factory Buildings* (New York: Industrial Extension Institute, 1919), 235; George M. Brill, "Location, Arrangement, and Construction of Manufacturing Plants," *Journal of the Western Society of Engineers*, 4 March 1908, 8; and Herbert Gilkey, Glenn Murphy, and Elmer O. Bergman, *Materials Testing* (New York: McGraw-Hill, 1941), 84.

2. I offer a revision of the views of Alfred Chandler, who excepts construction from the early-twentieth-century mass-production movement. See Alfred Chandler, *The Visible Hand: The Managerial Revolution in American Business* (Cambridge, Mass.: Harvard University Press, 1971), 242. Also Arthur Stinchcombe, "Bureaucratic and Craft Administration of Production: A Comparative Study," *Administrative Science Quarterly* 4 (1959): 169.

3. Marc Silver's 1986 study of work in the contemporary building trades offers a corrective to the exceptionalism that histories of American production have attached to construction. See Marc Silver, *Under Construction: Work and Alienation in the Building Trades* (Albany: SUNY Press, 1986), 1–13.

4. William Haber, *Industrial Relations in the Building Industry* (1930; reprinted New York: Arno and New York Times, 1971), 36.

5. Ernest Ransome, "New Developments in Unit Work Using a Structural Concrete Frame and Poured Slab," *Cement Age* 12 (March 1911): 130.

6. For an extensive description of American factory design in the nineteenth century, see Lindy Biggs, *The Rational Factory* (Baltimore: Johns Hopkins University Press, 1996). Also Charles K. Hyde, "Assembly-Line Architecture: Albert Kahn

and the Evolution of the U.S. Auto Factory, 1905–1940," *IA: The Journal of the Society for Industrial Archeology* 22 (1996): 5–24.

7. On early manufacturing plants in the United States, see John Phillips Coolidge, *Mill and Mansion: A Study of Architecture and Society in Lowell, Massachusetts, 1820–1865* (New York: Columbia University Press, 1942); Betsy W. Bahr, "New England Mill Engineering: Rationalization and Reform in Textile Mill Design, 1790–1920" (Ph.D. diss., University of Delaware, 1988); Betsy H. Bradley, *The Works: The Industrial Architecture of the United States* (New York: Oxford University Press, 1999); James F. Munce, *Industrial Architecture: An Analysis of International Building Practice* (New York: F. W. Dodge, 1960); Daniel Nelson, "The Factory Environment," chap. 2 in *Managers and Workers: Origins of the New Factory System in the United States, 1880–1920* (Madison: University of Wisconsin Press, 1975); and Judith A. McGaw, *Most Wonderful Machine: Mechanization and Social Change in Berkshire Paper Making, 1801–1885* (Princeton: Princeton University Press, 1987).

8. Siegfried Giedion, *Space, Time, and Architecture* (1945; reprinted Cambridge, Mass.: Harvard University Press, 1967), 184; Nelson, *Managers and Workers,* 12. Reyner Banham suggests that the tall, narrow form of early American industrial buildings may have been derived from industrialists' second- or thirdhand knowledge of the dockside warehouses of European port cities. These buildings were the tallest and narrowest of the early-eighteenth-century cityscapes because they had been designed to make maximum use of expensive real estate, and because vertical cartage by crane was less expensive than haulage over horizontal distances. Reyner Banham, *A Concrete Atlantis: U.S. Industrial Building and European Modern Architecture, 1900–1925* (Cambridge, Mass.: MIT Press, 1986), 40.

9. Cast-iron-skeleton buildings did have a lasting popularity in commercial loft construction taking place in American cities in the mid-nineteenth century. Built on smaller lots on crowded streets, these buildings frequently had cast-iron and glass facades of some distinction. Most notable in this area was the work of James Bogardus in New York between 1850 and 1870. See Munce, *Industrial Architecture,* 4–5, and Giedion, *Space, Time, and Architecture,* 195–99.

10. Leonard Eaton, "The Gateway City and Its Warehouses," in *The Gateway City and Other Essays* (Ames: Iowa State University Press, 1989), 10.

11. Nelson, *Managers and Workers,* 13. See also Bahr, *The Works.*

12. For examples of the varied types of building required by some manufacturing processes, see McGaw, *Most Wonderful Machine,* 224, 227–28. Also Gail Cooper, "The Seasonal Factory in the United States, 1900–1919," unpublished ms., 1992, and *Air-Conditioning America* (Baltimore: Johns Hopkins University Press, 1998).

13. Nelson, *Managers and Workers,* 11–13. Another impetus to the erection of standardized industrial facilities after 1870 was the burgeoning distribution and merchandizing side of manufacturing. As Leonard Eaton has written, mercantile capitalism relied on vast stores of goods held at the ready for national markets.

Warehouses were erected to hold these goods. Often conveniently centralized in industrial neighborhoods, the buildings were generally of simple and undifferentiated design, with the exception of elaborate showrooms added to some. Eaton, "Gateway City," 4.

14. Case, *Factory Buildings,* 264–66. For a comprehensive discussion of the nature of interior factory design, see Biggs, *Rational Factory.*

15. Brill, "Location, Arrangement, and Construction," 3.

16. The sawtooth roof was an English invention that first spread to the United States in the 1870s. Its period of greatest popularity in this country was from 1900 to 1910. Fred S. Hinds, "Saw Tooth Skylight Factory Roof Construction," *Transactions of the American Society of Mechanical Engineers* 28 (1906): 35, cited in Nelson, *Managers and Workers,* 13.

17. Carl Condit refers to Turner's system as "the first mature technique of column-and-slab framing." The reinforcing for the mushroom system combined concentric, radial, and continuous multiple-way rods to counter the complicated stresses that occurred where a slab rested on a column. Carl W. Condit, *American Building Art: The Twentieth Century* (New York: Oxford University Press, 1961), 167–68.

18. Case, *Factory Buildings,* 331–35; Clayton Mayers, *Economy in the Design of Reinforced Concrete Buildings* (Boston: Aberthaw Construction Co., 1918), 26. The economy of flat-slab construction derived also from the fact that its use of simpler forms meant that wood need not be cut into small pieces and therefore could be reused. "Barton Spider Web System," *Sweet's Catalog,* 1914, 220.

19. Haber, *Industrial Relations in the Building Industry,* 21. By 1926, domestic Portland cement production had reached 164 million barrels. For representative discussions of the advantageous nature of reinforced-concrete factory construction, see *The Library of Factory Management,* vol. 1, *Buildings and Upkeep* (Chicago: A. W. Shaw Co., 1915), 52–76, and H. Colin Campbell, "Building the Factory of Concrete," *Industrial Management,* March 1917, 806–16.

20. On the material complexities of large-scale food and drug manufacture, see Cooper, *Air-Conditioning America.*

21. Banham, *Concrete Atlantis,* 63.

22. Charles Day, *Industrial Plants: Their Arrangement and Construction* (New York: Engineering Magazine, 1911), 57.

23. Concrete also had other disadvantages as a material for factory buildings. In the first few years of the twentieth century, concrete buildings suffered from a certain type of inflexibility. Shafting and belting were still commonly used for power transmission; electricity was not yet established. Once a reinforced-concrete building was erected, it was difficult to alter the arrangement of shafting. Holes could not be easily made in concrete beams or floors, and new fixtures could not be easily added. Moreover, before 1910, engineers had not yet solved the

difficulties of building with concrete in cold weather. Concrete construction was therefore confined to the warm months in which labor was in the greatest demand and thus the most expensive.

24. Banham, in *Concrete Atlantis,* offers a lengthy aesthetic analysis of the design of reinforced-concrete factory buildings in the United States.

25. Ransome (1844–1917) was the son of Frederick Ransome, who had contributed to the development of the British concrete industry, most notably through the creation of a rotary cement kiln. Ernest Ransome began his career in California in the 1870s addressing the difficulty of building to resist earthquakes. In 1885 he constructed a flour mill with floor slabs cast integrally with supporting beams, each beam reinforced in its tension zone with a single rod. By 1889 he had developed a homogenous system of floor construction in which girders, beams, and slabs of a given bay were cast as a unit on concrete columns. Ernest L. Ransome, "Reminiscence," in *Reinforced Concrete Buildings* (New York: McGraw-Hill, 1912), reprinted in *A Selection of Historic American Papers on Concrete, 1876–1926,* ed. Howard Newlon Jr. (Detroit: American Concrete Institute, 1976), 291–93; Henry J. Cowan, *The Master Builders,* vol. 1, *Science and Building: Structural and Environmental Design in the Nineteenth and Twentieth Centuries* (New York: John Wiley and Sons, 1977), 36, 79.

26. Banham, *Concrete Atlantis,* 70.

27. Albert Kahn (1869–1942) established his practice in Detroit. In addition to industrial commissions, he worked on residential, institutional, and civic buildings. Julius Kahn (1874–1942) founded the Kahn Trussed Concrete Steel Company in Detroit and established affiliates of this business around the world while working closely with his brother. For information on the Kahns, see Grant Hildebrand, *Designing for Industry: The Architecture of Albert Kahn* (Cambridge, Mass.: MIT Press, 1974); Biggs, *Rational Factory;* Terry Smith, *Making the Modern* (Chicago: University of Chicago Press, 1993); Peter Conn, *The Divided Mind: Ideology and Imagination in America, 1898–1917* (Cambridge: Cambridge University Press, 1983); and Federico Bucci, *Albert Kahn: Architect of Ford* (New York: Princeton Architectural Press, 1993).

28. Day, *Industrial Plants,* 79–80; Harold V. Coes, "Better Industrial Plants for Less Money," *Factory: The Magazine of Management* 17 (January 1917): 22.

29. Lockwood, Greene's history is related in Samuel Lincoln Bicknell, *Lockwood Greene: The History of an Engineering Business, 1832–1958* (Brattleboro, Vt.: Stephen Greene Press, 1960). For a discussion of the aesthetic merits of some of the company's early-twentieth-century buildings, see Banham, *Concrete Atlantis,* 85–89.

30. Stone and Webster, "Industrial Plants and General Building—Work Done and Personnel Available, Vols. I and II," Special Collections of the Baker Library of the Harvard University Graduate School of Business Administration.

31. Haber, *Industrial Relations in the Building Industry*, 58.

32. Ross F. Tucker, "The Progress and Logical Design of Reinforced Concrete," *Concrete Age* 3 (September 1906): 333.

33. See advertisements for Turner Construction Company in *Sweet's Catalog*, 1907–8, 153, and Standard Concrete-Steel Company, same volume, 154.

34. Walter Mueller, "Reinforced Concrete Construction," *Cement Age* 3 (October 1906): 321–32; A. J. Widmer, "Reinforced Concrete Construction," *Twenty-fifth Annual Report of the Illinois Society of Engineers and Surveyors*, 1915, 148.

35. The Hinchman-Renton Company was based in Denver, which may be why it considered barbed wire to be "inexpensive and readily obtained in any quantity." Mueller, "Reinforced Concrete Construction," 325.

36. Advertisement for Clinton Wire Cloth Company, *Sweet's Catalog*, 1906, 96.

37. Mueller, "Reinforced Concrete Construction," 323–24; advertisement for Unit Concrete Steel Frame Company, *Sweet's Catalog*, 1906, 127.

38. The Aberthaw Construction Company had purchased patent rights to Ransome's twisted-steel reinforcement designs in 1896 and manufactured reinforcing from them for its own use and for sale. When the patent rights expired, probably around 1915, Aberthaw found it uneconomical to produce its own rods given the many "deformed" rods then on the market, especially those of the Kahn Company. See "The Story of Aberthaw," unpublished ms., ACC Archives, n.p.

39. Ransome, "New Developments in Unit Work," 129–34; J. L. Peterson, "History and Development of Precast Concrete in the United States," *JACI* 25 (February 1954): 477–96. Peterson credits John E. Conzelman with the development of the first system of precast elements to be applied to industrial buildings on a nationwide scale. Conzelman assembled fifty-one patents between 1910 and 1916 for elements of his system. Ibid., 484.

40. Advertisement for Turner Construction Company, *Sweet's Catalog*, 1907–8, 153.

41. John Sewell, U.S. Army Corps of Engineers, complained of this restraint in an article on the latest construction methods: "Unfortunately, the use of attached stirrups is patented in this country." John Sewell, "Reinforced Concrete," *Cement Age* 1 (November 1904): 206.

42. Advertisement for John W. Allison Company, Philadelphia, *Sweet's Catalog*, 1907–8, 135. See also advertisement for Concrete Steel Company, New York, same volume, 137, and advertisement for Tucker and Vinton Company, New York, *Sweet's Catalog*, 1906, 126.

43. "Ransome Concrete Machines—1908 Models," trade catalog in Architectural Trade Catalog Collection, Avery Architectural Library, Columbia University.

44. George Underwood, *Standard Construction Methods* (New York: McGraw-Hill, 1931), 19. For a description of representative mixing machines as of 1920, see Nathan C. Johnson, "Concrete Equipment," in *Handbook of Building Construction*,

vol. 2, ed. George A. Hool and Nathan C. Johnson (New York: McGraw-Hill, 1920), 862–64. In a move not then common for manufacturers of large machines, Ransome produced the mixers in "lots of ten or twelve" rather than on order, to keep costs low.

45. Haber, *Industrial Relations in the Building Industry,* 28–29; Ransome, "New Developments," 133–34.

46. Hool and Johnson, *Handbook of Building Construction,* 2:867–69. For a detailed description of one extremely elaborate building site, see George P. Carver, "Reinforced Concrete Building Work for the United Shoe Machinery Company, Beverly, Massachusetts," *Engineering News* 53 (25 May 1905): 537–40.

47. Mayers, *Economy in the Design,* 11.

48. Frank E. Kidder and Harry Parker, *Kidder-Parker Architects' and Builders' Handbook* (New York: John Wiley and Sons, 1931), 1241.

49. Underwood, *Standard Construction Methods,* 18, 128.

50. Chandler, *Visible Hand,* 241–42.

51. Chandler's exegesis of the history of American corporate management can be found in ibid. and in *Strategy and Structure: Chapters in the History of the Industrial Enterprise* (Cambridge, Mass.: MIT Press, 1962). Chandler finds the origins of these managerial methods in the organizational schemes of mid-nineteenth-century American railroads. Charles O'Connell locates them in the operations of the United States Army of the same period. See Charles F. O'Connell Jr., "The Corps of Engineers and the Rise of Modern Management, 1827–1856," in *Technological Change: Perspectives on the American Experience* (Cambridge, Mass.: MIT Press, 1985).

52. Haber, *Industrial Relations in the Building Industry,* 59.

53. Haber also claims that smaller firms were prone to getting involved in "unsound" competitive practices and that for this reason building trade unions preferred to work with the large integrated building firms. Ibid., 61, 72.

54. Advertisement for Barton Spider Web System, *Sweet's Catalog,* 1920, 220.

55. Brill, "Location, Arrangement, and Construction," 3.

56. Day, *Industrial Plants,* 18.

57. Frank D. Chase, *A Better Way to Build Your New Plant* (Chicago: Frank D. Chase, 1919), 4.

58. Mayers, *Economy in the Design,* 5.

59. Ibid., 6.

60. Widmer, "Reinforced Concrete Construction," 148. Widmer advocated the Kahn System by name. This raises the question of whether engineers had a certain fidelity to a given system and whether there was any mutual commitment on the part of the fabricating company. See Amy Slaton and Janet Abbate, "The Hidden Lives of Standards," in *Technologies of Power,* ed. Gabrielle Hecht and Michael Allen (Cambridge, Mass.: MIT Press, forthcoming).

61. Day, *Industrial Plants*, 96.

62. See, for example, advertisement for Roebling Construction Company, *Sweet's Catalog*, 1907-8, 148. Ironically, the concrete engineers' message was reinforced by other building trades that felt threatened by the introduction of systemized building methods. The president of the Bricklayers', Masons', and Plasterers' International Union said in 1905, "Concrete construction is a dangerous undertaking and requires the most skilled and intelligent direction. Our organization has contended that the adoption of the various concrete systems used in construction are experimental and uncertain at best, and the work requires the most skillful mechanics." President Bowen, *Bulletin*, Building Trades Association, New York, September 1905, 228, cited in Haber, *Industrial Relations in the Building Industry*, 39.

63. Case, *Factory Buildings*, 254.

64. Day, *Industrial Plants*, 4; Case, *Factory Buildings* 255. On the role of generalized knowledge in the creation of modern professions, see Theodore Porter, *Trust in Numbers* (Princeton: Princeton University Press, 1995).

65. Harry C. Bates, *Bricklayers' Century of Craftsmanship* (Washington, D.C.: Bricklayers, Masons, and Plasterers International Union of America, 1955), 79.

66. "Public Getting Wise to Wood Substitutes," *Carpenter*, April 1916, 42.

67. Henry Sterling Chapin, "A Revolution in Building Materials," Building Brick Association of America, brochure, 1911, 10; Bates, *Bricklayers' Century*, 139, 143.

68. Bates, *Bricklayers' Century*, 146.

69. President Bowen, cited in ibid., 144.

70. The Aberthaw Construction Company shrank considerably over the second half of the twentieth century. It was dissolved entirely in the late 1980s, although its name is still used by a small firm located in Billerica, Massachusetts. Records and original promotional materials dating from 1902 are held by this office (referred to here as ACC Archives). Additional published brochures are maintained in the Avery Architectural Library of Columbia University and various university engineering history collections.

71. Recollections of Edward Temple in "The Story of Aberthaw," unpublished ms., Aberthaw Construction Company, ACC Archives, n.p. Thompson's methods are described in Sanford E. Thompson and William O. Lichtner, "Scientific Methods in Construction," *Proceedings of the Engineers Society of Western Pennsylvania*, 1916, 434-65.

72. On the early history of the Aberthaw Construction Company, see ibid. and "Reinforced Concrete as Used in Manufacturing Plants," brochure, 1902, ACC Archives.

73. Aberthaw Construction Company, "For the Leaders," catalog, 1915, ACC Archives.

74. "Aberthaw Construction Service," brochure, 1918, ACC Archives, 5. Contemporary prescriptive texts recommend to construction site managers that there always be work for every class of laborer: excavation for the common worker, form building for carpenters, and so on. See Underwood, *Standard Construction Methods,* 126. The attention paid to this seemingly obvious means of speeding up and economizing building suggests the degree to which post-1900 construction departed from earlier, craft-based techniques. Many more workers were present on the building site, and the intimate relations of the apprentice-master system no longer controlled the division or pace of labor.

75. "Aberthaw Construction Service," 7; "Construction Service, Open-Shop Principles, and Organization and Personnel of the Aberthaw Company," ms., ca. 1925, ACC Archives, 3. Aberthaw employed many workers for multiple projects, some staying on the payroll for years at a time. This was a relatively rare procedure in the construction industry made possible here by Aberthaw's size.

76. Chandler, *Visible Hand,* 240–44.

77. Ibid., 272–74.

78. On the role of internal communications in systematic management methods, see JoAnne Yates, *Control through Communication: The Rise of System in American Management* (Baltimore: Johns Hopkins University Press, 1989).

79. "Story of Aberthaw," n.p.

80. Ibid.

81. "40 Days at Colt's Arms," brochure, 1917 or 1918, ACC Archives, 10, 5.

82. See Chandler, *Visible Hand,* 240–83.

83. "40 Days at Colt's Arms," 10, 13; "Story of Aberthaw," n.p.; "Instruction Booklet," unpublished ms., 1931, ACC Archives, n.p.

84. "Story of Aberthaw," n.p.

85. "40 Days at Colt's Arms," 7.

86. "Aberthaw Construction Service," 15–16.

87. Daniel Nelson, *Managers and Workers,* 52–53. Henry R. Towne of Yale and Towne is credited with developing the first gain-sharing plan in an effort to improve plant operation. Unlike premium or profit-sharing plans, the 50 percent gain-sharing program saved owners money equal to that paid out in bonuses, and Aberthaw felt that the system had appeal to its clients for this reason. "Story of Aberthaw," n.p.

88. "Story of Aberthaw," n.p. Evidence that the bonus was intended as a special reward for workers is provided in Edward Temple's recollection of Aberthaw's payment system: "The regular wages were paid in the ordinary manila pay envelope, while the bonus payment was in a red envelope. The regular pay envelope went home to the wife, while the red envelope became the workmen's personal money." Ibid. Similar sentiments about workers' home lives appear in the literature regarding labor issues that Aberthaw prepared for clients. Workers' wives were

scapegoated as causing labor discontent: "Domestic Pressure Aids the Strike." See Morton C. Tuttle, "The Housing Problem in Its Relation to the Contentment of Labor," brochure, Aberthaw Construction Company, 1920, ACC Archives. On incentives, see David Montgomery, *Workers' Control in America: Studies in the History of Work, Technology, and Labor Struggles* (Cambridge: Cambridge University Press, 1979), 114–17.

89. Nelson, *Managers and Workers,* 51–53.
90. "Construction Service, Open-Shop Principles," 18.
91. "Instruction Booklet," n.p.
92. Haber, *Industrial Relations in the Building Industry,* 72.
93. Ibid., 72–76. Charles Day expressed this sentiment in his 1911 text on factory construction: "The only feasible bases upon which to remunerate an engineering organization for such service are a fixed fee or percentage of the actual cost, for the economies effected through the competitive purchase of all materials just at the time they are needed, directly from those who can quote the lowest prices, should accrue to the owner. . . . The [building] services rendered by the engineering organization should be considered in no wise less professional than the work previously performed by them, resulting in the plans and specifications." Day, *Industrial Plants,* 95–96. See also Daniel J. Hauer, "Contracts," in *Handbook of Building Construction,* 1069–70.
94. Advertisement for the Underwriters Engineering and Construction Company, *Sweet's Catalog,* 1907–8, 146.
95. Edward Temple outlined Aberthaw's contracting policy in his recollections. As general manager in the 1920s, he had been asked, "How can anyone expect a builder to exert the same effort and skill in saving every dollar he can in the building cost when his profit is not at stake (as it is in a lump-sum bid)?" He explained that only by avoiding any hint of diseconomy or impropriety could the cost-plus-fee firm continue to operate. By definition, it seems, Aberthaw was a high-integrity firm, and no artificial incentives such as shaving costs or quality of work would figure in its operations. "Story of Aberthaw," n.p. In essence, Temple claimed that his firm's work was above reproach and the need for inspection: "I have often heard Architects and Engineers refer to their careful inspection which results in having work done properly. I have often wondered if they think they could inspect Ted Williams into hitting a home-run it he did not want to, or if it was to his money advantage not to do so." Ibid.
96. "Instruction Booklet," n.p.
97. "Story of Aberthaw," n.p.
98. See especially "Seamless," a brochure documenting Aberthaw's construction of a New Haven rubber plant in 1919 and 1920, ACC Archives. The text and photographs emphasize the rationalized production methods of the Seamless Rubber Company and equate Aberthaw's methods with its client's. Also of inter-

est are "For the Leaders—Built by Aberthaw," 1919; "Four Aberthaw Clients Move Ahead Again," ca. 1919; and "Building for Industrial Leaders," 1925 (all ACC Archives).

99. Stinchcombe, "Bureaucratic and Craft Administration," 170.

100. *Sweet's Catalog,* 1906, 172.

CHAPTER FIVE. What "Modern" Meant

Epigraph: G. H. Edgell, *The American Architecture of To-Day* (New York: Charles Scribner's Sons, 1928), 287–88.

1. Architectural journals of 1900 to 1920 frequently published articles on the necessity of increasing the role of architects in factory design. See "The Architect and the Industrial Plant," *American Architect* 113 (27 February 1918): 250–58, and Arthur J. McEntee, "Recent Development in the Architectural Treatment of Concrete Industrial Buildings," *Architecture* 43 (January 1921): 18–21.

2. Representative discussions by factory-building firms and specialists include W. Fred Dolke Jr., "Some Essentials in the Construction of an Industrial Building," *American Architect* 3 (21 February 1917): 116, and Ross F. Tucker, "The Progress and Logical Design of Reinforced Concrete," *Cement Age* 3 (September 1906): 233.

3. Reyner Banham, *A Concrete Atlantis: U.S. Industrial Building and European Modern Architecture, 1900–1925* (Cambridge, Mass.: MIT Press, 1986); Terry Smith, *Making the Modern: Industry, Art, and Design in America* (Chicago: University of Chicago Press, 1993), 57–92.

4. Richard Wightman Fox and T. J. Jackson Lears, eds., *The Culture of Consumption: Critical Essays in American History, 1880–1980* (New York: Pantheon, 1983); Peter Conn, *The Divided Mind: Ideology and Imagination in America, 1898–1917* (Cambridge: Cambridge University Press, 1983).

5. These phrases appear in "Utilitarian Structures and Their Architectural Treatment," *American Architect* 96 (10 November 1909): 183. The sentiments are echoed throughout contemporary discussions of factory design in architectural and technical journals and in books on factory operation.

6. Ibid., 186.

7. Dolke, "Some Essentials," 116.

8. Tucker, "Progress and Logical Design," 233.

9. This doctrine emerged in the writings of some architects and critics in the last quarter of the nineteenth century. Some took the position of Louis Sullivan (for whom the "organic" nature of materials was the appropriate source of design "truth"). Others followed the reasoning of Ruskin that the nature of craftwork itself should determine design. Peter Conn ably summarizes these tendencies in relation to functionalist industrial architecture. See Conn, *Divided Mind,* 210–22.

I am identifying a proliferation of truth-to-materials rhetoric after 1900 with particular reference to concrete.

10. "Announcement," *American Architect* 91 (4 May 1907): 161.

11. A. D. F. Hamlin, "The Architectural Problem of Concrete," *American Architect* 91 (4 May 1907): 163. Similarly, in the same issue, "From Messrs. Horgan & Slattery, New York, NY," 167; "From Mr. Morrison H. Vail, New York, NY," 171; Evarts Tracy, "The Idiomatic Use of Concrete," 172; and E. P. Goodrich, "Concrete Walls and Floors," 177.

12. H. Toler Boorem, "Architectural Expression in a New Material," *Architectural Record* 23 (April 1908): 265. Another architect in the *American Architect's* May 1907 survey wrote, "The ideal of decoration would seem to be to give a pleasing appearance to each member or portion of a structure and in such a way as at the same time to disclose its real use and composition. If this ideal is sound, the employment of such details as imitation beams formed in plaster under large-span concrete slabs is not good architecture." E. P. Goodrich, 177.

13. Russell Sturgis, "The Warehouse and Factory in America," *Architectural Record* 15 (January 1904): 14. In 1906 Sturgis wrote even more explicitly that "he is no true student of architecture who does not love bricks and stones for themselves—for their weight, their permanent squareness, their sharp edged, flat-bedded quality—the warehouse-and-factory way of design is peculiarly susceptible to this means of expression." Russell Sturgis, "Factories and Warehouses," *Architectural Record* 19 (May 1906): 369. Conn describes more generally the context of "realistic design" in which Sturgis and other critics practiced, and discusses the work of Albert Kahn. Kahn's factories are famously austere, in variance with his residential, academic, and civil commissions, which frequently bore very traditional detailing. Kahn stands outside this study because he was trained as an architect and illuminates the cutting edge of factory design rather than its dissemination, but he here exemplifies the belief that different kinds of building were thought to call for radically different architectural styles. Conn, *Divided Mind,* 214–19.

14. C. Canby Balderston, Victor S. Karabasz, and Robert P. Brecht, *Management of an Enterprise* (New York: Prentice-Hall, 1935), 98.

15. Thorstein Veblen, *Theory of the Leisure Class* (1899; reprinted New York: A. M. Kelley, 1965), 110–18, cited in Conn, *Divided Mind,* 201.

16. "Utilitarian Structures," 183.

17. McEntee, "Recent Development," 18.

18. "Utilitarian Structures," 186; *The Library of Factory Management,* vol. 1, *Buildings and Upkeep* (Chicago: A. W. Shaw Co., 1915), 87.

19. Warren R. Briggs, "Modern American Factories," *Architecture* 38 (September 1918): 231.

20. *Library of Factory Management,* 87.

21. Stuart D. Brandes, *American Welfare Capitalism, 1880–1940* (Chicago: University of Chicago Press, 1970); David Brody, *Workers in Industrial America: Essays on the Twentieth Century Struggle* (Oxford: Oxford University Press, 1980), 51–69; Michael B. Katz, *In the Shadow of the Poorhouse* (New York: Basic Books, 1986), 187–90; David Noble, *America by Design: Science, Technology, and the Rise of Corporate Capitalism* (New York: Oxford University Press, 1977), 286–89.

22. Lindy Biggs, *The Rational Factory* (Baltimore: Johns Hopkins University Press, 1996); Gail Cooper, *Air-Conditioning America* (Baltimore: Johns Hopkins University Press, 1998).

23. For representative discussions of plant improvements, see Portland Cement Association, "Mercantile and Industrial Buildings of Concrete," *Architect and Engineer* 62 (July 1920): 81; Ernest G. W. Souster, *The Design of Factory and Industrial Buildings* (London: Scott, Greenwood, and Son, 1919), 34–40, 69–89; and the letters of the industrialist Henry S. Dennison in the Baker Library of the Harvard Graduate School of Business Administration. Dennison visited the factory of the National Cash Register corporation in Dayton, Ohio, in 1900 and recorded his observations. Of working conditions, he wrote, "The presence of plants in crepe-covered pots and the simple vines on the rafters of the girls lunch room gave a simple, natural beauty to the place; one fern in a room full of benches served to ornament the whole section. Yet on the face of every article of this sort was written plainly: 'Economy—cheap and good.' Not one cent's worth of tinsel was to be seen. The upper side of a chair seat was painted white, the lower side was unpainted." Papers of Henry S. Dennison, File: "1900–1915," Special Collections, Harvard Business School.

24. Aberthaw Construction Company, "A Report on Reconstruction and Relocation for the Seamless Rubber Company, Inc.," unpublished ms., 1919, ACC Archives, 23; Aberthaw Construction Company, "Seamless: How Close Knit Cooperation Developed a Rubber Factory a Quarter Century in Advance of Its Time," brochure, ca. 1920, ACC Archives, n.p.

25. "Aberthaw Construction Services," 1918, ACC Archives, 13.

26. Leifur Magnusson and Robert L. Davison, "Housing by Employers in the United States," *Bulletin of the Bureau of Labor Statistics* 263 (1920): 9.

27. Given that Harvard's "social ethicists" were also working in this period on studies of such subjects as divorce and temperance, it would appear that worker housing had joined these social phenomena as a matter of social scientific concern. See David B. Potts, "Social Ethics at Harvard, 1881–1931," in *Social Sciences at Harvard, 1860–1920,* ed. Paul Buck (Cambridge, Mass.: Harvard University Press, 1965), 97, 119.

28. Reports also included ethnicities of workers and whether they were born in the United States, held foreign citizenship, or were naturalized citizens. Aberthaw believed that these facts correlated with its clients' desire for ready and depend-

able workers. See, for example, Aberthaw's "Report on Reconstruction and Relocation for the Seamless Rubber Company." Aberthaw's housing experts may also have believed in correlations between workers' housing arrangements and job performance such as those suggested in the Bureau of Labor Statistics study. Authors of the 1916 study proposed that married workers who owned their own homes were less likely to leave their jobs, and that the presence of unrelated boarders in a home could lead to degeneration of a stable family setting. Magnusson and Davison, "Housing by Employers," 15–18.

29. Morton Tuttle, "The Housing Problem in Its Relation to the Contentment of Labor," brochure, Aberthaw Construction Company, 1920, ACC Archives.

30. Ibid., 3.

31. The authors of Aberthaw's report concluded, "In short it would seem that under existing conditions the Rubber Company Plant tends to attract only a somewhat restricted class of labor, and that, of this class the women remain only during what may be called the period of indecision, up to 25 years of age, whereas the men come only after they have spent their earlier years being sifted through other forms of employment." Aberthaw Construction Company, "A Report on the Reconstruction of the Seamless Rubber Plant, New Haven, Connecticut," 1919, ACC Archives, 23–25.

32. The circumscribed role that Aberthaw envisioned for American factory workers found expression in the public image the company crafted for itself from 1905 onward. In the language and illustrations of its promotional materials, Aberthaw lent itself an aura of vital importance for the growth of American industry. Fliers from around 1905 depict various manufacturing firms "Moving Ahead With Aberthaw." A 1915 catalog of several dozen Aberthaw buildings equates each with improved client production. In addition to photographs of Aberthaw's finished buildings, many of the publications include dramatic illustrations of half-completed buildings rising on handsome waterfronts or before sweeping cityscapes. The place of Aberthaw's construction workers in this scenario of industrial expansion is conveyed by the inclusion of muscular, usually faceless construction workers wielding shovels or operating cranes. The men appear to be a sturdy, orderly, and anonymous element of the building process, at once machinelike and heroic. The illustrations offer a mixed message that outwardly venerates manual labor while subsuming individual workers' identity or experience under the larger purposes of capitalism, echoing in many ways the company's own operations and its recommendations to its industrial clients.

33. Willard Case, *The Factory Buildings* (New York: Industrial Extension Institute, 1919), 258.

34. See Stephen Meyer III, *The Five Dollar Day: Labor Management and Social Control in the Ford Motor Company, 1908–1921* (Albany: State University of New York Press, 1981), chap. 7, "Assembly-Line Americanization."

35. Descriptions are based on "Scranton Yards of the Delaware, Lackawanna, & Western Railroad," report, Historic American Engineering Record, National Park Service, U.S. Department of the Interior, HAER-PA-132-K, 1993. Significant contemporary articles include "The New Locomotive Repair Shops of the Lackawanna Railroad at Scranton, Pa.," *Railway Age* 43 (12 April 1907): 597–601, and George L. Fowler, "Scranton Shops of the Delaware, Lackawanna & Western," *Railroad Age Gazette* 42 (5 November 1909): 865–72. See also Amy Slaton, "Aesthetics of a Modern Industry: Buildings of the Delaware, Lackawanna, and Western Railroad's Scranton Yards," *IA: Journal of the Society for Industrial Archeology* 22 (1996): 25–39.

36. The DL&W ceased operations in Scranton in the 1960s. The site has recently been developed as a national park, and most of the buildings discussed still stand.

Conclusion

Epigraph: Charles P. Stivers, "Concrete" (poem), *Cement Era* 12 (1914): cover.

1. Owen B. Maginnis, "Exactness in Carpentry and Joinery," *Carpenter,* April 1916, 6–7.

2. Advertisement, *Carpenter,* June 1914, 60.

3. Ibid., April 1914, 59.

4. Frank Duffy, "Choosing an Occupation," *Carpenter,* April 1916, 5; "Our Principles," *Carpenter,* April 1914, 21; Harry C. Bates, *Bricklayers' Century of Craftsmanship* (Washington, D.C.: Bricklayers, Masons, and Plasterers International Union of America, 1955), 108.

5. The complexity of understanding worker radicalism and its absence in America is conveyed in David Montgomery, *The Fall of the House of Labor* (Cambridge: Cambridge University Press, 1987).

Bibliographic Essay

Primary Sources

The development of concrete for commercial use after 1900 was a heterogeneous task. Experts and entrepreneurs developed and implemented new physical entities (materials, machines, instruments) and new knowledge systems (data, protocols, and managerial techniques). For scientists, engineers, and owners of commercial building firms, all of these technical tasks were paired with the additional labor of establishing and critiquing reputations. To account for as many of these activities as possible, I have used the records of individuals and institutions operating in a great variety of settings.

The most significant primary sources in this study are the standards and specifications for concrete issued by the American Society for Testing Materials, the American Society of Civil Engineers, and cement trade organizations as a basis for contracts and other legal instruments of the building industry. The *Proceedings* of the ASTM and the ASCE, as well as publications of the Portland Cement Association (Skokie, Ill.) and the American Concrete Institute (Detroit), present these written instruments in many iterations from about 1895 onward. The standards and specifications appear as well in innumerable commercial contracts, engineering textbooks, and civil codes. To consult almost any legal or educational discussion of concrete construction from 1900 onward is to encounter some version of these protocols.

Because it offers a "life story" of such instruments—following them from their academic and industrial origins out through their daily use in the building trades—this book begins with the work of university engineering departments that specialized in the study of concrete. I examined most closely the archives of engineering departments at the University of Illinois (Champaign-Urbana), the University of Pennsylvania, and Iowa State University. All have maintained excellent records of their work on modern construction materials, in part because leading engineers in these schools saw themselves as public figures, serving industry and the larger polity through science. In addition to carefully preserving correspondence and teaching materials, these three universities have made Arthur Talbot, Edwin Marburg, and Anson Marston, respectively, subjects of much commemorative attention over the last century; celebratory narratives of telling heft abound. Smaller schools that did early and interesting work in the testing of commercial concrete, such as Lehigh University, Drexel University, and the former

Lewis Institute of Chicago (eventually subsumed by the Illinois Institute of Technology), have unfortunately kept fewer records, but their activities are reflected in individual school bulletins and histories and, most important, in the various publications of the Society for the Promotion of Engineering Education.

SPEE was founded in 1894, and in the years under study here it produced a wide variety of proceedings, bulletins, and special publications. The history of the society and the complex system by which its publications were titled and numbered are recounted in Matthew Elias Zaret, "An Historical Study of the Development of the American Society for Engineering Education" (Ph.D. diss., New York University 1967); see also Terry Reynolds and Bruce Seely, "Striving for Balance: A Hundred Years of ASEE," *Engineering Education* 82 (July 1993): 136–51. Articles, transactions, and papers issued by the society illuminate a complex set of occupational and class relations surrounding the rise of the engineering professions, as do the hundreds of engineering textbooks and advice manuals published in this period. Extremely helpful background information on engineering education in the United States between 1900 and 1930, against which the history of materials testing can be understood, is offered in David Noble, *America by Design: Science, Technology, and the Rise of Corporate Capitalism* (New York: Oxford University Press, 1977), and Peter Lundgren, "Engineering Education in Europe and the U.S.A., 1750–1930: The Rise to Dominance of School Culture and the Engineering Professions," *Annals of Science* 47 (1990): 33–75.

To probe the experiences of work on the early-twentieth-century construction site and to determine how closely the prescriptions of the concrete experts and promoters were actually followed, I turned to the records of building firms of the era. Helpful records on the largest commercial construction firms of the day—including Stone and Webster and Lockwood, Greene—are represented in the collections of the Baker Library of the Harvard University Graduate School of Business Administration and are the subjects of laudatory but informative self-published histories. The Baker Library also offers selected blueprints and correspondence of firms that commissioned reinforced-concrete factory buildings in the early twentieth century, such as the Boston Woven Hose Company, and several projects of the Turner Construction Company. The lesser-known concrete building firms on which this study centers were far more difficult to trace. A few, such as the Foundation Company, can be studied through papers held in the National Museum of American History in Washington, D.C., which also holds the papers of Robert Cummings, a structural engineer who built many concrete factories. The Aberthaw Construction Company has a great many uncatalogued papers in storage in its offices in Billerica, Massachusetts. Additional information on companies that supplied cement and steel reinforcing, that specialized in erecting utility buildings, or that otherwise made their way in this burgeoning trade can be gleaned from advertising materials. *Sweet's Catalog* and the trade catalog collections of the Avery Architectural Library of Columbia University and the Hagley Museum and Library in Wilmington, Delaware, are invaluable in this regard. Promotional materials of the Portland Cement Association and the American Concrete Institute are also accessible through those organizations and in the collections formerly held by the Engineering Societies Library of New York, now largely maintained by the Linda Hall Library in Kansas City, Missouri.

The history presented here of the instruments with which cement and concrete were tested after 1900 is based on several sources. There is little written history on materials testing as a scientific or commercial undertaking: Stephen P. Timoshenko's *History of Strength of Materials: With a Brief Account of the History of Theory of Elasticity and Theory of Structures* (New York: Dover Publications, 1953) provides needed background for the catalogs and narrative publications produced by the major instrument makers early in the century, including those of the Riehle, Tineus Olsen, and Baldwin-Southwark companies (maintained in the Linda Hall Library). Invoices, correspondence, and catalogs of these and other instrument firms that specialized in cement and concrete equipment also appear in the papers of many university engineering instructors. Periodicals of the cement industry—*Cement Age, Concrete,* and *Rock Products* all began publication shortly after 1900—contain both advertisements and editorial coverage of testing apparatus. An exceptionally interesting source for this book was the instrumentation installed in the Owyhee Dam in Owyhee, Oregon, a structure built by the U.S. Bureau of Reclamation in the 1920s as a large-scale test site for technologies used in the construction of the Hoover Dam. Gauges and meters, and even clipboards and rulers, used to measure the strength and heat of curing mass concrete remain in place and present an application of the testing and inspection methods being taught in universities in this period.

Finally, I consider the actual factory buildings discussed here to be primary sources of paramount importance. Many stand along the Northeast Corridor of the former Pennsylvania Railroad (now Amtrak) and on the edges of old downtowns from Los Angeles to Chicago to Baltimore. To walk around and through them is to understand something of the scale of effort required of their designers and builders. Some buildings that have been subjected to renovation or adaptive reuse are well documented by local historic-landmark commissions. Buildings of the Delaware, Lackawanna, and Western Railroad, described in chapter 5, are documented in Historic American Engineering Record, National Park Service, U.S. Department of the Interior, HAER-PA-132-K, 1993.

Secondary Sources

With its emphasis on the social origins and consequences of modern concrete construction, this book intentionally subverts distinctions between science and engineering as intellectual projects of a "basic" and "applied" nature, respectively. Nor does it seek to categorize as "experimental" or "practical" episodes in the development and use of concrete technologies. Instead, I treat the work of people who studied, tested, and built with concrete after 1900 as *work*—activity undertaken for purposes of employment, self-promotion, or the general economic welfare of some company or citizenry. Through such an approach there emerge commonalities of class, gender, and economic ambition across occupations and institutions. To understand as socially comparable the open-ended researches of materials experts and the routine quality-control tasks of field-testers and to associate these enterprises with the managerial agendas of building firm managers, I turn to a range of historical models. Labor history, the history of technology, the sociology of professions, and the literature of science studies have all supplied crucial grounding for this study.

The rise of expert professions in the Progressive Era is well documented. This book has been inspired by the generation of historians who sought explanations for the emergence of scientific and social-scientific fields after 1900 in the broadest societal realignments of this period, rather than in any self-evident triumph of scientific reason over political self-interest. The bases of this approach are found in Samuel P. Hays, *Conservation and the Gospel of Efficiency: The Progressive Conservation Movement, 1890–1920* (Cambridge: Harvard University Press, 1959) and *The Response to Industrialism, 1885–1914* (Chicago: University of Chicago Press, 1957), and developed in Robert Wiebe's *The Search for Order, 1877–1920* (New York: Hill and Wang, 1967). Wiebe's more recent *Self-Rule: A Cultural History of American Democracy* (Chicago: University of Chicago Press, 1995) is also suggestive. However, as rich as these studies are in crafting portraits of technical experts as social, rather than strictly intellectual, actors, they in some ways cast Progressive Era reformers as reactive—as if some ambient disturbance naturally moves people to take organizational command in times of disruption. Alternatively, the exceedingly proactive engineers of David Noble's *America by Design* display a range of motivations—from gender and race bias to status concerns—that may explain why some Americans of means developed occupations based on technical expertise. Noble explicates the social conservatism of certain Progressive engineers in a way that Edwin Layton's otherwise informative *Revolt of the Engineers* (1971; reprinted Baltimore: Johns Hopkins University Press, 1986), elides. For discussion of the coexistence of "business-friendly" and reformist attitudes among American engineers after 1900, see Peter Meiksins, "The 'Revolt of the Engineers' Reconsidered," *Technology and Culture* 29 (1988): 219–46.

BIBLIOGRAPHIC ESSAY

The problem of how science achieved its reputation as a politically neutral undertaking after 1900 forms a core concern of this book. That science-based quality control so often bore an aura of neutrality while serving specific commercial or occupational interests suggests that disinterest is an intellectual category, at least in an industrial age, of some complexity. We are faced with the question of how particular bodies of knowledge acquired social authority in a given setting: What characteristics do experts appropriate to make their new knowledge meaningful to tradition-bound audiences? The classic histories of the major engineering societies—Daniel Calhoun, *The American Civil Engineer: Origins and Conflict* (Cambridge, Mass.: Technology Press, MIT; distributed by Harvard University Press, 1960), and Monte A. Calvert, *The Mechanical Engineer in America, 1830–1910: Professional Cultures in Conflict* (Baltimore: Johns Hopkins Press, 1967)—were written before such questions gained currency in the history of technology. Bruce Sinclair, in *A Centennial History of the American Society of Mechanical Engineers, 1880–1980* (Toronto: University of Toronto Press, 1980), begins to ask the vital question of what problems professionalizing engineers chose to address as they sought to distinguish their work from that of "lesser" technical occupations. The epistemologies of modern engineering fields were constructed with jurisdictional aims in mind, and evidence of "disinterest" served important reputational functions. The Progressive Era, as Weibe saw, gave rise to many cases in which information itself served to construct group identities. Two case studies from outside the history of technology offer some guidelines on how such constructions occurred: JoAnne Brown, *The Definition of a Profession: The Authority of Metaphor in the History of Intelligence Testing, 1890–1930* (Princeton: Princeton University Press, 1992), and Jonathan L. Zimmerman, *Distilling Democracy: Alcohol Education in America's Public Schools, 1880–1925* (Lawrence: University of Kansas Press, 1999). To delve deeply into the success of engineering as both social and industrial force after 1900, we might combine these sociologically informed approaches with the methodologies of science studies. Here we find ways to associate the precise nature of testing and inspection work on the construction site with the social advantages it conferred on its promoters.

That the social identity of elite practitioners brings credibility to their practice is a well-developed idea in the history of science. Foundational works include Steven Shapin, "The House of Experiment in Seventeenth-Century England," *Isis* 79 (1988): 373–404, and "Pump and Circumstance: Robert Boyle's Literary Technology," *Social Studies of Science* 14 (1984): 481–520; and Steven Shapin and Simon Schaffer, *Leviathan and the Air-Pump: Hobbes, Boyle, and the Experimental Life* (Princeton: Princeton University Press, 1985). On the ways in which the control of material and social resources perpetuates intellectual authority in scientific settings, see Robert Kohler's *Lords of the Fly: Drosophila Genetics and the Experimental Life* (Chicago: University of Chicago Press, 1994). The question remains how these compelling historical formulations can be imported into the study of materials testing and inspection—uses of information seemingly so routine as to defy analysis as part of an intellectual hierarchy. I take cues here from historians of science who have already "pushed the envelope." In their essay on historiographic approaches to the field sciences, Henrika Kuklick and Robert Kohler ("Introduction," *Science in the Field,* ed. Henrika Kuklick and Robert Kohler, *Osiris* 11 [1996]: 1–14) make clear that the practices, physical conditions, and social identities of scientific occupations are mutually determinative in settings well beyond the laboratory or clinic. Intentionally or otherwise, they invoke some of the emphasis that labor history has placed on working conditions and structures of opportunity and risk in work—of which science and engineering are, of course, examples. Emphasizing technical fields, Philip Scranton surveys progress in this interdisciplinary effort in recent decades in "None-Too-Porous Boundaries: Labor History and the History of Technology," *Technology and Culture* 29 (1988): 722–43. I treat concrete construction in exactly this way: the nature of testing protocols, the material and labor conditions of the construction site, and the competitive pressures experienced by all who work in the building trades, from scientists on "down," are mutually formative and must be studied together if any feature of this enterprise is to be thoroughly understood.

Intimately related are issues of intellectual neutrality as a historically contingent ideal of scientific inquiry, and here work on objectivity in science is very helpful. The idea that the exercise of judgment carries a shifting valuation in the scientific occupations—at times more or less valued than "objective" practice—is developed in Lorraine Daston and Peter L. Galison, "The Image of Objectivity," *Representations* 40 (1992): 81–128, and Peter L. Galison, "Judgment against Objectivity," in *Picturing Science, Presenting Art,* ed. Peter L. Galison and Caroline A. Jones (London: Routledge, 1998), 327–59. The persistence of subjective practice in advanced technical work is evident in the case of reinforced concrete and is further explained with the help of sociological studies of professions by Andrew Abbott, *The System of Professions: An Essay on the Division of Expert Labor* (Chicago: University of Chicago Press, 1988), and Everett Hughes, *The Sociological Eye* (Chicago: Aldine Atherton, 1971), in which the success of occupations is seen to be a multiply determined entity. All measures of "talent" aside, if a profession is to thrive, the nature of services offered must be subject to specialization; research questions must be chosen on the grounds that they are eligible for resolution but not so readily that amateurs may answer them first. Essays in Ronald G. Walters, ed., *Scientific Authority and Twentieth-Century America* (Baltimore: Johns Hopkins University Press, 1997), and Thomas Haskell, ed., *The Authority of Experts* (Bloomington: Indiana University Press, 1984), are especially helpful in tracing this pragmatism across a range of scientific and social-scientific enterprises. In many of these essays about emerging bodies of expertise, occupational programs are linked to issues of class, race, and ethnicity in ways that inform my own study of materials testing after 1900.

As engaging and helpful as this literature on the development of elite epistemologies is, it leaves out the sort of intellectual and social maneuvering to which materials experts and their clientele applied themselves as participants in routine technical undertakings. Standardized technical labor—the development and implementation of protocols for technical work—has lately received increasing attention from historians and sociologists of science, and it is from these works that I have borrowed descriptive methodologies. Of greatest interest are Stefan Timmermans and Marc Berg, "Standardization in Action: Achieving Local Universality through Medical Protocols," *Social Studies of Science* 47 (1997): 273–305; Linda F. Hogle, "Standardization across Non-standard Domains: The Case of Organ Procurement," *Science, Technology, and Human Values* 20 (1995): 482–500; Geoffrey C. Bowker and Susan Leigh Star, *Sorting Things Out: Classification and Its Consequences* (Cambridge, Mass.: MIT Press, 1999); Warwick Anderson, "The Reasoning of the Strongest: The Polemics of Skill and Science in Medical Diagnosis," *Social Studies of Science* 22 (1992): 653–84; and Karen Rader, "'The Mouse People': Murine Genetics Work at the Bussey Institution, 1909–1936," *Journal of the History of Biology* 31 (1998): 327–54. One of the few works on standardization to address engineering practice is Stuart Shapiro, "Degrees of Freedom: The Interaction of Standards of Practice and Engineering Judgment," *Science, Technology, and Human Values* 22 (1997): 286–316, but Shapiro tends to focus on features of standardized design methods as they are engendered within the engineering workplace rather than as a function of larger social or class contestation. A valuable survey of the commercial origins and uses of materials standards is Samuel Krislov, *How Nations Choose Product Standards and Standards Change Nations* (Pittsburgh: University of Pittsburgh Press, 1997). Krislov undertakes some analysis of the ideological features of modern standardization movements but misses the broader social impacts of these enterprises; Noble's *America by Design* remains a much more provocative and helpful analysis.

The idea that one can locate systems of social privilege through studying the epistemological features of quality control remains a new subject for historians that no doubt awaits the breakdown of longstanding disciplinary boundaries, but one work has provided a particularly salient model for this book: Ken Alder's *Engineering the Revolution: Arms and Enlightenment in France, 1763–1815* (Princeton: Princeton University Press, 1997), in which commercial pressures, distributions of technical skills, and political ideologies of the greatest philosophical reach weave inseparably through the history of industrial society. Alder problematizes the "ownership" of quality—who determines and enforces what is "good enough" in the world of production—in a way that can be translated to later and much larger scale examples. For almost all businesses of the last two centuries, the maintenance of quality and uniformity in production has been a matter of constant concern, and it is surprising that materials testing, and in fact technical testing of any description, has received so little attention from historians. Bruce Seely's *Building the American Highway System* (Philadelphia: Temple University Press, 1987) remains by far the most complete description of the testing and inspection activities of the Progressive Era and is a remarkably provocative account of professional and institutional ambition in an influential group of engineers.

This study seeks to connect such ambitions to the actual nature of materials testing and inspection—to the intellectual features of testing. Only a few studies have laid the groundwork for such a connection. In his brief sociological analysis of scientific and technological testing, Trevor Pinch outlines the importance for testers of establishing the veracity of the test itself—how closely it resembles the conditions under which a material will be used, for example. Pinch here offers an important analytical tool for probing the negotiated nature of testing and inspection ("'Testing—One, Two, Three . . . Testing': Toward a Sociology of Testing," *Science, Technology, and Human Values* 18 [1993]: 25–41). In his sociological consideration of testing, Donald Mackenzie broadens the social context in which the usefulness of a given test is established to include military and political forces; see "The Construction of Technical Facts," chap. 7 in *Inventing Accuracy: An Historical Sociology of Nuclear Missile Guidance* (Cambridge, Mass.: MIT Press, 1990).

We might consider the labor of testing as a subset of measuring and sorting, and here another group of scholars offers guidance. These are the historians of science who have chosen to problematize order within the world of commerce, tracking scientific methods that variously shape or confront the operations of business: Simon Schaffer, "Accurate Measurement Is an English Science" (135–72), and Graeme J. N. Gooday, "The Morals of Energy Metering: Constructing and Deconstructing the Precision of the Victorian Electrical Engineer's Ammeter and Voltmeter" (239–82), both in *The Values of Precision,* ed. M. Norton Wise (Princeton: Princeton University Press, 1995); Theodore Porter, *Trust in Numbers* (Princeton: Princeton University Press, 1995); and Geoffrey Bowker, *Science on the Run: Information Management and Industrial Geophysics at Schlumberger, 1920–1940* (Cambridge, Mass.: MIT Press, 1994). What is particularly pleasing about these works is that no unit of classification or measurement is deemed so mundane that it cannot delegate labor, economic risk, or social status. Scientific constants, accounting systems, and cement sieves all do in fact emerge from intentional organizations of human effort. JoAnne Yates, *Control through Communication: The Rise of System in American Management* (Baltimore: Johns Hopkins University Press, 1989), and Olivier Zunz, *Making America Corporate, 1870–1920* (Chicago: University of Chicago Press, 1990), complement these works by melding business administration systems and the social organization of offices in suggestive ways.

BIBLIOGRAPHIC ESSAY

The problem of achieving uniformity in productive settings is of course richly described in modern labor history and the history of technology, and I need not repeat the leading titles in the history of mass production, Fordism, and Taylorism. The reader would best begin with David Hounshell, *From the American System to Mass Production, 1800 to 1932* (Baltimore: Johns Hopkins University Press, 1984); Harry Braverman, *Labor and Monopoly Capital: The Degradation of Work in the Twentieth Century* (New York: Monthly Review Press, 1974); David Montgomery, *Workers' Control in America: Studies in the History of Work, Technology, and Labor Struggles* (Cambridge: Cambridge University Press, 1979); Daniel Nelson, *Managers and Workers: Origins of the New Factory System in the United States, 1880–1920* (Madison: University of Wisconsin Press, 1975); and David Gordon, Richard Edwards, and Michael Reich, *Segmented Work, Divided Workers* (Cambridge: Cambridge University Press, 1982). All profile the emergence of mass-production and managerial methods after 1900 and their variable impacts on managers and workers.

Importantly, a number of authors in recent years have shown that even within "flow-based" enterprises, idiosyncrasy persists. This might be in the form of "custom" design (Philip Scranton, *Proprietary Capitalism: The Textile Manufacture at Philadelphia, 1800–1885* [Cambridge: Cambridge University Press, 1983], and *Endless Novelty: Specialty Production and American Industrialization, 1865–1925* [Princeton: Princeton University Press, 1997]; Gail Cooper, *Air-Conditioning America* [Baltimore: Johns Hopkins University Press, 1998]) or in the negotiations and intuitions exercised on the shop floor (Michael Nuwer, "From Batch to Flow: Production Technology and Work-Force Skills in the Steel Industry, 1880–1920," *Technology and Culture* 29 [1988]: 808–38; Charles Sabel and Jonathan Zeitlin, "Historical Alternatives to Mass Production: Politics, Markets, and Technology in Nineteenth Century Industrialization," *Past and Present* 108 [1986]: 173–96). These "deviations" from flow were not impediments to large-scale production, and they help illuminate the persistence of nonstandardized knowledge in the testing and inspection of concrete. Finally, William Cronon's marvelous connection of the material features of modern production (the liquidity of grain, the propensity of lumber to warp) and its administrative methods (systems of pricing and grading) in *Nature's Metropolis* (New York: W. W. Norton, 1991) has been a crucial model for my linkage of concrete and the technical protocols that made it a viable building medium.

The project of treating building as a technological and cultural enterprise can lead one into two historiographic traps: that which treats building as unlike other industrial operations because it relies on craft knowledge; and that which treats building as unlike other productive undertakings because it has so many associations with the "high-art" traditions of architecture. Concrete fits neither stereotype, and this book is in fact an attempt to undermine those preconceptions about building.

BIBLIOGRAPHIC ESSAY

Many of the best secondary sources on the technical features of early-twentieth-century concrete come from the preservation field. Important bibliographies are E. L. Kemp, ed., *History of Concrete: 30 B.C. to 1926 A.D., Annotated Bibliography No. 14* (Detroit: American Concrete Institute, 1982), and U.S Department of the Interior, National Park Service, Preservation Assistance Division, "Twentieth Century Building Materials, 1900–1950" (Washington, D.C., 1993). For a general historical overview of the material, Cecil Elliott's *Technics and Architecture: The Development of Materials and Systems for Building* (Cambridge, Mass.: MIT Press, 1992) augments Henry J. Cowan's *The Master Builders,* vol. 1, *Science and Building: Structural and Environmental Design in the Nineteenth and Twentieth Centuries* (New York: John Wiley and Sons, 1977). *Reinforced Concrete up to 1914,* ed. Frank Newby (Brookfield, Vt.: Ashgate Publishing, forthcoming), promises helpful stylistic and technical coverage as well.

The building trades—here I use the phrase to refer to both managerial and laboring sectors of the industry—are not widely studied by historians. This book does not attempt to report fully on the experiences of the lesser-trained workers on the construction site whose experiences might be best sought in the records of the builders' unions. Instead, with something of a "supply-side" emphasis, I focus on the conditions perceived and manipulated by owners of building firms and those who provided materials and services to those firms. For this purpose, William Haber's *Industrial Relations in the Building Industry* (1930; reprinted New York: Arno and New York Times, 1971) proved a comprehensive if somewhat dated source. Gwendolyn Wright, *Moralism and the Model Home: Domestic Architecture and Cultural Conflict, 1873–1913* (Chicago: University of Chicago Press, 1980), remains exceptional in its resolute integration of the experiences of building designers, firm owners, and laborers and its connection of all three to the aesthetics of modern home design in 1900. Mark Silver, *Under Construction: Work and Alienation in the Building Trades* (Albany: SUNY Press, 1986), is extremely important, particularly for correcting the longstanding notion that building is exceptional in the history of modernizing industries. Alfred Chandler, *The Visible Hand: The Managerial Revolution in American Business* (Cambridge, Mass.: Harvard University Press, 1977), describes building as resembling mining and agricultural processes in relying on traditional skills and the use of hand tools, and in having few aspects that could be subject to speeding up or to an intensified use of energy. While it is true that building may have been relatively labor-intensive when compared with highly mechanized manufacturing enterprises ("where machinery replaced men"), concrete construction showed a number of the characteristics Chandler ascribes to mass-production contexts. We find in the reinforced-concrete building firm of 1910 "a complex organization to coordinate the flow of goods from one process to another." Silver demonstrates that in particular cases, construction throughout the twentieth century has displayed as substantial a degree of manager control and worker alienation as other industries.

In approaching reinforced-concrete factory buildings as the product of new labor relations and technical developments after 1900, this book attempts to restructure arguments about the origins of modernist design in the United States. Nonetheless, existing work on factory design provides important descriptive material, and one might best begin with the classic by John Coolidge, *Mill and Mansion: A Study of Architecture and Society in Lowell, Massachusetts, 1820–1865* (New York: Columbia University Press, 1942). With much more explicit attention to the relationship of technologies and skills employed *within* the factory to the design of factories, Judith McGaw describes the form of early American paper mills in *Most Wonderful Machine: Mechanization and Social Change in Berkshire Paper Making, 1801–1885* (Princeton: Princeton University Press, 1987). Additional extremely helpful descriptions of factory design that refer to plant operation appear in Lindy Biggs, *The Rational Factory* (Baltimore: Johns Hopkins University Press, 1996), and Betsy H. Bradley, *The Works: The Industrial Architecture of the United States* (New York: Oxford University Press, 1999). Monographs on the most widely known factory designer of the early twentieth century come from Grant Hildebrand, *Designing for Industry: The Architecture of Albert Kahn* (Cambridge, Mass.: MIT Press, 1974), and Federico Bucci, *Albert Kahn: Architect of Ford* (Princeton: Princeton Architectural Press, 1993).

Arguably the most influential aesthetically focused study remains Reyner Banham's *A Concrete Atlantis: U.S. Industrial Building and European Modern Architecture, 1900–1925* (Cambridge, Mass.: MIT Press, 1986). Banham's book drew the attention of architectural historians to a long neglected but commonplace feature of the American landscape and celebrated its sophisticated visual nature and historical importance to modernist European design. His emphasis was on the work of Kahn and other aesthetic innovators. My own study on virtually anonymous building designers and members of other construction occupations locates enduring cultural influences in the *least* exceptional examples of concrete factory buildings and denies the primacy of European innovators in the history of American modernism, but in many ways Banham's book helped legitimize my own choice of subject.

There is a great deal of literature on the American embrace of novelty—in design, in lifestyle, in intellectual enterprise—after 1900, and much of it has aided this study of architectural change. For a portrait of architectural modernism with appealingly broad cultural and social connections, see David Ward and Olivier Zunz, eds., *The Landscape of Modernity: Essays on NYC, 1900–1940* (New York: Russell Sage Foundation, 1992), especially the essays by Keith Revell and Marc Weiss, which embed modern city planning and building design in systems of entrenched political power. Emily Thompson's dissertation, "'Mysteries of the Acoustic': Architectural Acoustics in America, 1800–1932" (Princeton University, 1992) injects into the story of modernist building design the history of another body of scientific knowledge and expertise—acoustics—to provide what I believe is a confirming case of technical, cultural, and occupational ambition in Progressive Era building. What these studies do best, I find, is acknowledge a mixture of the new and old in American work and culture of the period. Peter Conn, in *The Divided Mind: Ideology and Imagination in America, 1898–1917* (Cambridge: Cambridge University Press, 1983), depicts such a blend. By *divided* I do not take Conn to mean an ambivalence about change often perceived by critics in America's combination of traditional and modern design forms in the twentieth century. Rather, *amalgam* may be a more suitable term for the nature of modernist culture in the United States. In concrete buildings of 1900 to 1930, at least, there is evidence of a modernity made up of new knowledge and old social orders: cutting-edge science enacted old-fashioned social hierarchies through its rejection of artisanal skills. Few studies of modernist design connect visual change to the systems of opportunities and risks emerging in people's daily work lives in the new century, and a social history of design is what is most needed if we are to understand how landscapes change.

Index

Aberthaw Construction Company (*see also* factory buildings; Mayers, Clayton; Temple, Edward H.; Tuttle, Morton): labor union relations, 157–63; operations, 127, 153, 157–67, 172, 178–83, 187, 229n. 71, 235n. 32, 238–39; public image, ftspc., 163–66, 172

Abrams, Duff, 85, 210n. 16, 217n. 21

Academy of Mines, Freiburg, Germany, 37

acceptance tests, 27

ACI. *See* American Concrete Institute

acoustics, architectural, 248

Adams, Ralph, 40, 43, 49

aesthetics, architectural (*see also* design), 13, 168, 170, 174–76, 183, 232n. 9, 247–48

AFL. *See* American Federation of Labor

American Architect, 174

American Brotherhood of Cement Workers, 156

American Concrete Institute (ACI; *see also* National Association of Cement Users), 28, 32, 74, 82, 98, 111, 237, 239

American Federation of Labor (AFL), 156, 157

American Ice Company (concrete-curing room), 37

American Institute of Architects, 99, 120

American Institute of Consulting Engineers, 120

American Institute of Electrical Engineers, 122, 216n. 14

American Railway Engineering Society, 68

American Sheet and Tin Plate Company, 167

American Society of Civil Engineers (ASCE; *see also* American Society for Testing Materials), 11, 31, 55, 65, 68, 80–82, 98–99, 104, 111, 237; Committee on Uniform Methods of Test, 81; Joint Committee on Concrete and Reinforced Concrete, 31; Joint Conference on Uniform Methods of Tests and Standardized Specifications for Cement, 99

American Society for Engineering Education. *See* Society for the Promotion of Engineering Education (SPEE)

American Society of Mechanical Engineers, 101, 216n. 14, 217n. 16

American Society of Safety Engineers, 220n. 51

American Society for Testing Materials (ASTM; *see also* American Society of Civil Engineers), 11, 62, 65, 111, 119, 125, 126, 210n. 14, 237; academics' role within, 31, 32, 65, 79; history, 28, 68; industrial objections to, 69, 114, 123–24; Joint Committee on Concrete and Reinforced Concrete, 31, 99; Joint Conference on Uniform Methods of Tests and Standard Specifications for Cement, 99; standards and specifications, 68–71, 81–90, 96, 98–102, 118

American Steel and Iron Company, 180

American Sugar Refinery, 140

Architectural Record, 175

Architecture, 177

Army Corps of Engineers, U.S., 69, 90, 99, 125, 210n. 11

ASCE. *See* American Society of Civil Engineers
Association of American Portland Cement Manufacturers (*see also* Portland Cement Association), 86, 96, 102–3
Association of Railway Maintenance of Way Engineers, 99
ASTM. *See* American Society for Testing Materials

Baldwin-Southwark Company, 239
Bell Laboratory, 110
Bell Telephone, 108
Berry, H. C., 35, 42, 108–10; portable strain gauge, 78, 79, 219n. 36
Bessemer process, 27, 29
Boston Woven Hose Company, 238
brand names. *See* contracts, brand names in
Breckenridge, Lester Paige, 107
Brickbuilder, 122
Bricklayers and Masons International Union (B&MIU), 156, 157, 229n. 62
Bridgeport Brass Company, 158
Briggs, Warren, 177
British Board of Trade, 118
Brown, Charles C., 122
Building Brick Association of America, 156
building codes: fire codes: —London, 117; —New York, 117, 221n. 74; specifications in, 116–21, 222n. 70; structural codes, 117, 119–20; —Chicago, 119–20 (*see also* McArdle, Peter)
Bureau of Labor Statistics, U.S., 180
Bureau of Public Roads, U.S., 111

Cameron Septic Tank Company, 109
Carnegie Foundation, 45, 60
Carpenters Union. *See* United Brotherhood of Carpenters and Joiners of America
Carter Ink Company, 158

Case, Willard, 154, 182
Case School of Applied Science, 112
cement (*see also* Association of American Portland Cement Manufacturers; concrete; Portland Cement Association): annual U.S. production of, 18, 135; artificial, 15; definition of, 15; live loads, 217; natural, 15; Portland, 15, 18–19, 122
Cement Age, 74, 111–13, 115, 122, 143, 239
Chandler, Alfred, 151, 246–47
character and personal attributes, engineers' (*see also* engineering education; objectivity in engineering), 8, 50, 58–60
Civil War, 26, 29
Coignet, François, 199n. 17
Colt's Patent Fire Arms Manufacturing Company, 160–61
Columbian Exposition (Chicago, 1893), 30
Columbia University, 28, 54
Comer, Henry, 120
concrete (*see also* cement): block, 15, 18; definition, 15; failures, 212n. 29; fireproofing use, 15–16, 117; flow table, 40; Gilmore needle, 40; history, 15–19, 246; mixing machines: —Eureka Company, 147; —Ransome Company, 146; precast (unit systems; *see also* Ransome, Ernest), 17, 144–45, 157, 192, 227n. 39; "readi-mix" ("transit-mix"), 73, 192; reinforced (*see also* reinforcing systems, commercial), 13, 134–46; reinforced construction methods: —beam-and-girder method, 134–36; —flat-slab method, 134–36; —Turner's mushroom column, 153; Vicat needle, 40; water-cement ratio, 72–74, 85, 192, 210n. 16
Concrete, 239
Concrete Engineering, 112

INDEX

construction. *See* factory construction
consultants. *See* materials testing
contracts: brand names in, 86, 96, 113–16; specifications in, 12, 67–68, 70, 95–96, 119–21
Cornell University, 54
Cowan, Henry, 118
Cowdrey, Irving, 40, 43, 49
Cummings, Robert, 238

"daylight" factory, 17
Delaware, Lackawanna & Western Railroad (DL&W), 183–86, 239
Dennison, Henry S., 234n. 23
Department of Agriculture, U.S., 69
Department of Commerce and Labor, U.S., 115
design (*see also* factory buildings): functionalist, 168–77, 183–86, 232n. 9; 247–48; history of factory, 130–32
Dolke, W. Fred , Jr., 174
Draffin, Jasper O., 28
Drexel University, 237
Dudley, Charles B., 20, 44, 45, 52, 59, 110, 197n. 10

Ellis, Charles, 46–47
engineering, objectivity in. *See* objectivity in engineering
engineering education (*see also* ethnicity; gender; objectivity in engineering; *and under names of specific universities*): automatic and autographic testing machines, 39, 43, 44; cooperative programs, 29; development of student intuition and judgment, 41–48, 54–55; German model, 37–38; histories, 238; ideologies of social leadership (*see* character and attributes, engineers'; testers and inspectors, occupational status of); liberal arts, 52–54; "rule-of-thumb" vs. scientific approaches, 20–21, 24–25; theory within curricula, 24, 44–49;

208n. 60; trend toward greater control of student work, 38–41
Engineering News, 112, 113
Engineering Record, 83, 112
Erie Canal, 26
ethnicity (*see also* race), 8, 56–60, 182, 191, 234n. 28
Eureka Company, 147

facsimile specifications (*see also* standards and specifications), 114
factory buildings: American Steel and Iron, 180; American Sugar Refinery, 140; Boston Woven Hose Company, 238; Bridgeport Brass Company, 158; Carter Ink Company, 158; Colt's Patent Fire Arms Manufacturing, 160–61; Delaware, Lackawanna & Western Railroad, 183–86; Ford Automobile Company, 138; Hood Rubber Company, 158; Pacific Coast Borax, 118, 134–35, 221n. 70; Pacific Printworks, 158, 162; Packard Automobile Company, 138; Pierce-Arrow Motor Car Company, 158; Samuel Cabot Company, 14; Seamless Rubber Company, 178–79, 180, 182, 231n. 98; Squantum Destroyer Plant, 163; Standard Oil Company, 172; United Shoe Machine Company, 138, 145; U. S. Rubber Refinery, 140
factory construction (*see also* aesthetics, architectural; design): fireproofing, 131–32, 134–37, 221n. 67; history of, 130–34; humanization of, 179; lighting, 17, 134, 136; signage, 137, 186; standardization of, 2–3, 13, 133–41, 154, 158–59, 171–77, 224n. 13; types, 130–35; worker housing, 170, 177–82, 235n. 28; worker safety and health, 23, 170, 177–79, 181
fixity, scientific, 7, 97, 121–25
Flexol Company, 110
Ford Automobile Company, 138

252 INDEX

Foundation Company, 210n. 14, 238
Franklin Institute, Philadelphia, 26
Fritz, John, 30
Frost, Harwood, 112–13
Functionalism. *See* design, functionalist

gang molds (for concrete), 144, 145
gender (*see also* Wilson, Elmina), 8, 51, 56–57, 192
Geological Survey, U.S. (USGS), 99
Gilkey, Herbert J., 32–33, 57, 85, 106
Grand Central Terminal, New York City, 150

Haber, William, 130, 140, 151–52, 164
Harvard University: Department of Social Ethics, 180; stadium, 158
Hatch Act (1889), 26
Hatt, W. K. 41, 43
Hennebique, François, 16, 145
Honest Reinforced-Concrete Culvert Company, 110
Hood Rubber Company, 158
Hoover Dam, 239
housing, worker. *See* factory construction
Hughes, Everett, 49
humanization (of factory design), 179
Humboldt Manufacturing Company (portable testing apparatus), 78
Humphrey, Richard, 84, 106
Hyatt, Thaddeus, 17, 199n. 21

Illinois Institute of Technology, 238
Insurance Engineering Experiment Station, 220n. 51
insurance industry, 132–33
International Correspondence School, Scranton, Pennsylvania, 191
International Society of State and Municipal Building Commissions, 119
International Style, 171
Iowa Engineer, 33
Iowa State College (ISC; *see also Iowa Engineer*; Gilkey, Herbert J.; Marston,

Anson), 26, 31–33, 37, 39, 41–44, 52, 56, 85, 107, 212n. 24, 237; cement laboratory, 37; Department of Theoretical and Applied Mechanics, 32, 35; Engineering Hall, 32; laboratory students, 43; moist curing rooms, 33–35; technical bulletins, 32
iron construction, 131, 224n. 9

Jackson, John Price, 52, 55
John W. Allison Construction, 146
journals, trade. *See under names of specific journals*

Kahn, Albert, 17, 138, 226n. 27
Kahn, Julius, 17, 138, 226n. 27; reinforcing systems, 138, 143, 145–46, 228n. 60; Trussed Concrete Steel Company, 226
Kreuzpointner, Paul, 86

labor unions (*see also under names of specific unions*), 157–63, 192, 246
land grant universities, 26
Layton, Edwin, 20
Lehigh University, 29–30, 38, 237
Lesley, Robert W., 33, 102, 112, 114, 202n. 29
Lewis Institute, 52, 54, 59, 238
liberal arts. *See* engineering education
Lockwood, Greene Company, 140, 158, 238
Louisiana Purchase Exposition (St. Louis, 1904): Engineering Congress, 122; Model Testing Laboratory, 105; Portland Cement Association exhibit, 104–6

Maillart, Robert, 134
Mann, Charles Riborg, 45, 48, 52, 207n. 85
Marburg, Edwin, 35, 39, 61, 113, 237
Marvin, Frank, 54
Marston, Anson, 32–33, 42, 44, 50, 52,

55, 57, 58, 61, 105–6, 109, 218n. 34, 237
Massachusetts Institute of Technology, 37, 40–41, 43, 54
materials science. *See* materials testing; strength of materials
materials testing (*see also* American Society of Civil Engineers; American Society for Testing Materials; strength of materials), 76–89; automation of, 39–40; consultants in, 22, 106–11; H. C. Berry portable strain gauge, 78; history of, 22–35; Humboldt Manufacturing Company Portable Test Apparatus, 78; instruments and machines for (*see also* Riehle Company), 23, 31, 35, 36, 76–80, 106–11, 212n. 28, 239; mechanization of, 23, 90–91; slump test, 77; sociology of, 244; university departments, 28–35 (*see also under names of specific universities*)
Mayers, Clayton, 153
McArdle, Peter, 126
McCollough, Ernest, 55
McCready, Ernest, 88–89, 124
Mead, Daniel, 49, 57–58, 75, 84–85, 95, 115
measurement, history of, 244
mechanization: of building, 4, 90–91, 146–51, 246; of materials testing, 23, 90–91, 215n. 69
Military Academy, U.S. (West Point), 26, 28
mill construction, 131–32
modernism (*see also* aesthetics, architectural; design), 14, 169–77, 183–86
Monier, Josef, 199n. 17
Morrill Land Grant Act (1862), 26, 29
mortars, 15
Municipal Engineer, 122

National Association of Cement Users (*see also* American Concrete Institute), 122–23, 126

National Board of Fire Underwriters, 117–18
National Bureau of Standards, U.S. (National Institute of Standards and Technology), 69, 217n. 15
neutrality, scientific. *See* objectivity in engineering
New York City subway system, 150
Nies, Frank, 184–86
Noble, David, 8

objectivity in engineering (*see also* personal equation), 7–8, 11–12, 87–90, 241
Office of Weights and Measures, Coast and Geodetic Survey, U.S., 69
Owyhee Dam, 239

Pacific Coast Borax, 118, 134–35; 221n. 70
Pacific Printworks, 158, 162
Packard Automobile Company, 138
Panama Canal, 19, 150
Pennsylvania Railroad, 27–28, 86; in-house laboratory, 27–28, 239
personal equation, 83–84, 214n. 48
Pierce-Arrow Motor Car Company, 158
Pond and Pond, 138
Portland cement. *See* cement, Portland
Portland Cement Association (*see also* Association of American Portland Cement manufacturers), 86, 96, 98, 102–7, 125, 126, 210n. 16, 217n. 21, 237, 239; *Editor's Reference Book*, 104; Model Testing Laboratory, 104
Potter, A. A., 58
Progressive Era, 240, 241, 244
Purcell and Elmslie, 138
Purdue University, 41, 43, 58

quality control, 4–9, 27–28, 64, 196, 241; in construction industry, 4, 9, 71, 74, 98–102, 141–43

race (*see also* ethnicity), 51, 56–60, 192
Ransome, Ernest, 17, 138, 144–46, 199n. 23, 226n. 25, 227n. 38; Ransome Company (concrete mixers), 146–47; Ransome Unit System, 144–45; United Shoe Machinery Company, 138, 145
regularity. *See* uniformity
reinforced concrete. *See* concrete, reinforced
reinforcing systems, commercial, 13, 142, 144–46
Rensselaer Polytechnic Institute, 26, 28, 50
Riehle Company, 35, 39, 239
Rock Products, 239
Roebling Construction Company, 229n. 62
Rosenberg, Charles, 108

safety, in construction (*see also* factory construction), 23, 71–76
Samuel Cabot Company, 14
sand library, 217
Schmidt, Garden, and Martin, 138
Scientific American, 111
Seamless Rubber Company, 178–80, 182, 231n. 98
sieves, 76, 77
Society of Automotive Engineers, 28
Society for the Promotion of Engineering Education (SPEE), 45, 48, 50, 53–54, 61, 238
Southwark-Emery Company, 36
Spackman, Henry, 114
specifications. *See* standards and specifications
Squantum Destroyer Plant, 163
Standard Concrete-Steel Company, 143
Standard Oil Company, 172
standards and specifications (*see also* American Society of Civil Engineers; American Society for Testing Materials), 64–71, 75, 80, 87–89, 94–126, 210nn. 11 & 14, 243; facsimile, 114; as managerial tool 9–11, 65–66, 84–87, 91–92; history of, 67–71; in interfirm contracts, 11, 95–98; for reinforcement, 99
steel production. *See* Bessemer process
Stevens Institute of Technology, 37–38
Stone and Webster, 140, 238
strength of materials (*see also* materials testing), 21–22, 28–35, 38
Sturgis, Russell, 175, 233n. 13
subjectivity. *See* objectivity in engineering; personal equation
Sweet's Catalog, 143, 239

Talbot, Arthur Newell, 30–33, 107, 237
Taylor, Frederick, 91, 158; *Principles of Scientific Management* (1911), 91
Taylor, W. Purvis, 86
Temple, Edward H., 160, 162, 165, 230n. 88, 231n. 95
testers and inspectors: education of (*see* engineering education); occupational status of, 3–6, 8, 10–12, 21, 23–24, 35–37, 52–55, 63–64, 241
textile mills, 131–32
theoretical and applied mechanics, departments of. *See* Iowa State College; University of Illinois
theory. *See* engineering education
Thompson, Sanford, 158
timber construction, 131–32, 135
Tinius Olsen Company, 35, 239
Towne, John Henry, 33
Towne Laboratory. *See* University of Pennsylvania
Towne Scientific School. *See* University of Pennsylvania
Trussed Concrete Steel Company (Kahn), 211n. 19
Tufts College, 28
Turner, C. A. P., 17, 134, 225n. 17

Turner Construction Company, 145, 238
Tuttle, Morton, 181

uniformity (as technical goal), 1–45, 76, 80–82, 137–38, 245
Unit Concrete Steel Frame Company, 144
United Brotherhood of Carpenters and Joiners of America, 155, 192
United Shoe Machine Company, 138, 145
U. S. Rubber Refinery, 140
University of Cincinnati, 29, 46; cooperative programs, 29
University of Illinois, 26, 28, 30–33, 38, 45, 85, 99, 107–8, 237: College of Engineering, 108, 219n. 40 (*see also* Talbot, Arthur Newell); Department of Theoretical and Applied Mechanics, 30, 31; Laboratory of Applied Mechanics, 31, 45
University of Iowa, 58
University of Pennsylvania, 33, 35, 39–42, 56, 79, 99, 108–110, 122, 206n. 75, 237; College of Engineering, 33; Department of Science (Towne Scientific School), 33; Lesley Cement Laboratory, 33; School of Engineering, 33; Towne Laboratory, 110; Wharton School, 175
University of Wisconsin, 75

Veblen, Thorstein, 175
Vicat, Louis-Joseph, 72
Vicat needle, 40

Waddell, J. A. L., 1, 50, 53
Ward, William, 16, 199n. 18; concrete house, 16
water-cement ratio. *See* concrete
Watertown Arsenal, 109, 122
Wickenden, William, 41; *Study of Engineering Education* (1930), 20, 48
Widmer, A. J., 153
Wilkinson, William, 199n. 17
Williams, C. C., 58, 59, 201n. 7
Wilson, Elmina (*see also* gender), 57, 208n. 97
Worcester Polytechnic Institute, 37
worker housing. *See* factory construction
worker safety and health. *See* factory construction
W. S. Tyler Company, 77